Algebraic Structures and Applications

PURE AND APPLIED MATHEMATICS

A Program of Monographs, Textbooks, and Lecture Notes

Contributions to *Lecture Notes in Pure and Applied Mathematics* are reproduced by direct photography of the author's typewritten manuscript. Potential authors are advised to submit preliminary manuscripts for review purposes. After acceptance, the author is responsible for preparing the final manuscript in camera-ready form, suitable for direct reproduction. Marcel Dekker, Inc. will furnish instructions to authors and special typing paper. Sample pages are reviewed and returned with our suggestions to assure quality control and the most attractive rendering of your manuscript. The publisher will also be happy to supervise and assist in all stages of the preparation of your camera-ready manuscript.

LECTURE NOTES

IN PURE AND APPLIED MATHEMATICS

Other Volumes in Preparation

Algebraic Structures and Applications

Proceedings of the First Western Australian Conference on Algebra

Edited by

Phillip Schultz
Cheryl E. Praeger
Robert P. Sullivan

Department of Mathematics
University of Western Australia
Nedlands, Australia

MARCEL DEKKER, INC.   New York and Basel

Library of Congress Cataloging in Publication Data

Western Australian Conference on Algebra (1st : 1980 :
    University of Western Australia)
    Algebraic structures and applications.

    (Lecture notes in pure and applied mathematics ; 74)
    1. Algebra--Congresses. 2. Groups, Theory of--
Congresses. I. Schultz, Phillip, [date].
II. Praeger, Cheryl E., [date].   III. Sullivan,
Robert P., [date].   IV. Title. V. Series.
QA150.W47 1980   512   81-22093
ISBN 0-8247-1570-5   AACR2

MARCEL DEKKER, INC.
270 Madison Avenue, New York, New York 10016

Current printing (last digit):
10 9 8 7 6 5 4 3 2 1

PRINTED IN THE UNITED STATES OF AMERICA

PREFACE

This volume contains twelve of the fifteen papers presented at the First
Western Australian Conference on Algebra, held in the Mathematics Department
of the University of Western Australia on June 6 through June 8, 1980.

The purpose of the Conference, which was organized by the editors of
this volume, was to bring together for fruitful discussion practitioners of
pure and applied algebra, and to celebrate in appropriate fashion the Golden
Jubilee of the Mathematics Department, which was founded in 1929.

The editors express their appreciation for the financial support of the
University of Western Australia, and of Trans Australia Airlines. Our thanks
also go to the able secretarial staff of the Mathematics Department, who prepared the final typescript; to the Media Services Department of the University, who were responsible for the artwork.

Phillip Schultz
Cheryl E. Praeger
Robert P. Sullivan

INTRODUCTION

In spite of the tendency of modern mathematics to fragment into ever more
specialized fields, there is a long tradition of the concepts and techniques
of one speciality being brought to bear on the outstanding problems of another, or on seemingly unrelated areas of the real world. Nowhere is this
more true than in Algebra, where in recent years we have seen brilliant applications to Physics, Chemistry, Communications and Economics.

The theme of the First Western Australian Algebra Conference was Algebra
and its Applications, and the papers presented there represent a diversity of
topics, some concerned with problems internal to their own branch of Algebra,
others with applications to other parts of Mathematics and Science.

The first four papers in this volume fall under the general rubric of
representations of finite groups, both in the classical sense and more con-
cretely as the group of automorphisms of symmetric graphs and as the group
of actions defined by the well-known Rubik's Cube.

Then follow a pair of papers on statistical applications, one using the
irreducible characters of finite abelian groups to generalize a classical
method in the design of experiments and the other using modular lattices,
and their Möbius functions, to diagonalize families of symmetric matrices
which arise in the theory of association schemes.

Next come two papers generalizing and extending known properties of
abelian groups to wider classes of algebras, in one case to modules over an
arbitrary commutative valuation ring, and in the other to radical classes
closed under direct products of various cardinalities.

The last four papers concern algebraic structures of various types. One
is a review of BCK-algebras, which are models of logical systems containing
only the implication functor. Another characterizes the Jacobson radical of
the endomorphism ring of subfree and injective valued vector spaces, structures used as a tool in the theory of abelian p-groups. There is a paper
dealing with problems concerning the least number of terms in a sum of products required to express an arbitrary member of a Grassmann space over various
fields. The final paper in the volume contains a characterization of the
ideal of semidistributive lattice varieties in a lattice of varieties.

CONTENTS

CONTRIBUTORS

R. A. BAILEY* Faculty of Mathematics, The Open University, Milton Keynes,
    England

WILLIAM H. CORNISH School of Mathematical Sciences, The Flinders University
    of South Australia, Bedford Park, South Australia, Australia

GARY DAVIS Department of Mathematics, La Trobe University, Bundoora,
    Victoria, Australia

L. FUCHS Department of Mathematics, Tulane University, New Orleans,
    Louisiana, U.S.A.

B. J. GARDNER Department of Mathematics, University of Tasmania, Hobart,
    Tasmania, Australia

I. MARTIN ISAACS[†] Department of Mathematics, University of Wisconsin,
    Madison, Wisconsin, U.S.A.

J. A. MacDOUGALL[††] Department of Mathematics and Computer Science, University
    of Prince Edward Island, Charlottetown, Prince Edward Island, Canada

CHERYL E. PRAEGER Department of Mathematics, University of Western Australia,
    Nedlands, Western Australia, Australia

HENRY ROSE School of Mathematical Sciences, The Flinders University of
    South Australia, Bedford Park, South Australia, Australia

CHRIS ROWLEY Faculty of Mathematics, The Open University, Milton Keynes,
    England

P. SCHULTZ Department of Mathematics, University of Western Australia,
    Nedlands, Western Australia, Australia

T. P. SPEED Department of Mathematics, University of Western Australia,
    Nedlands, Western Australia, Australia

---

*Current affiliation: Statistics Department, Rothamsted Experimental Station,
Harpenden, England

[†] At the time of writing the author was a Visiting Fellow at the Department
of Mathematics, I.A.S., Australian National University, Canberra, A.C.T.,
Australia

[††] At the time of writing the author was visiting the Department of Mathematics, The University of Newcastle, N.S.W., Australia

Algebraic Structures and Applications

NOTES ON ZERO DIVISORS AND GROUP PRESENTATIONS

GARY DAVIS

Department of Mathematics
La Trobe University
Bundoora, Victoria
Australia

1. THE ZERO DIVISOR QUESTION FOR GROUP RINGS

Let k be an integral domain and G be a group. When is the group ring kG
without zero divisors? If G has an element g of finite order n then
$1 + g + \cdots + g^{n-1}$ is a zero divisor in kG, so if kG has no zero divisors
then G is torsion-free. Is there a torsion-free group G whose group ring kG
has zero divisors, for some k? More specifically, is there a torsion-free
group whose integral group ring has zero divisors? An account of work on
this problem up to 1976 appears as chapter 13 of Passman's book [3].

The most significant result to this date is the theorem of Farkas and
Snider which asserts that if G is a torsion-free polycyclic-by-finite group
then $\mathbb{C}G$ has no zero divisors. The extension from polycyclic to polycyclic-
by-finite groups uses Euler characteristics and establishes that $\mathbb{C}G$ has a
division ring as a full ring of fractions [1].

2. A CONNECTION WITH FINITELY-PRESENTED GROUPS

Passman points out that if G is torsion-free then kG has a zero divisor if
and only if it has a nilpotent of index 2: this follows from the fact that
kG is a prime ring when G is torsion-free. The question of section 1 can
therefore be rephrased in the following way: is there a torsion-free group
G for which kG has a nilpotent of index 2, for some integral domain k (in
particular, for $k = \mathbb{Z}$)?

In the style of Higman's idea of unique products groups we can write
down a sufficient condition for kG to have no nilpotents of index 2. Consider the following possible property of a group G:

For each finite subset $\{x_1,\ldots,x_n\}$ of distinct elements of G some
$x_i x_j$ cannot be written in the form $x_k x_\ell$ with $i \neq k$ and $j \neq \ell$.   (*)

Clearly a group with property (*) has no nilpotents of index 2 in kG. We
therefore ask the following question: is there a torsion-free group that
does not have property (*)? A negative answer would settle the zero divisor
problem.

Suppose, indeed, that G is a torsion-free group for which property (*)
fails. Then for some finite subset $\{x_1,\ldots,x_n\}$ of G each $x_i x_j$ is equal to
some $x_k x_\ell$ with $i \neq k$ and $j \neq \ell$. Let H be the subgroup of G generated by
$\{x_1,\ldots,x_n\}$ and let P be the partition of $\underline{n}^2 = \{1,\ldots,n\} \times \{1,\ldots,n\}$ given
by: (i,j) and (k,ℓ) belong to the same P-class if and only if $x_i x_j = x_k x_\ell$.
Then H is a homomorphic image of the finitely-presented group

$$\Gamma = \langle t_1,\ldots,t_n \mid t_i t_j = t_k t_\ell \text{ if } (i,j) \text{ and } (k,\ell) \text{ lie in the same P-class} \rangle$$

Now a contradiction will ensue if in the group Γ we have $(t_i t_j^{-1})^K = 1$ for
some $i \neq j$, and for some integer K: due to the fact that the x_i are distinct
in the torsion-free group H.

Consequently we pose the following question: Let n be an integer >1 and
P be a partition of the set $\underline{n}^2 = \{1,\ldots,n\} \times \{1,\ldots,n\}$ such that

(a)   each P-class has cardinality >1

(b)   for each pair of elements $(i,j),(k,\ell)$ in a single P-class either
$i \neq j$ and $k \neq \ell$, or else $i = j$ and $k = \ell$.

Let Γ_P be the finitely presented group given by

$$\Gamma_P = \langle t_1,\ldots,t_n \mid t_i t_j = t_k t_\ell \text{ if } (i,j) \text{ and } (k,\ell) \text{ lie in the same P-class} \rangle$$

Is there a pair of integers i,j with $1 \leq i, j \leq n$ and $i \neq j$, and an integer
K, such that $(t_i t_j^{-1})^K = 1$ in Γ_P?

A positive answer to this question for all P would imply that all torsion free groups have the property (*) which would yield a negative answer to the questions of section 1; that is to say, resolve the zero divisor conjecture affirmatively.

3. A SPECIAL CASE

The groups Γ_P in which all (i,i) constitute a P-class are somewhat easier to handle than other groups of this form. This condition on P means that amongst the relations of Γ_P we have

$$t_1^2 = \cdots = t_n^2$$

Since t_1^2 is then central in Γ_P the subgroup $N = \langle t_1^2 \rangle$ is normal and $K_P = \Gamma_P/N$ is generated by involutions. The following lemma tells us that in this case we really can restrict our attention to the groups K_P rather than the Γ_P.

LEMMA. Let α_i denote the image of t_i in $K_P = \Gamma_P/N$. If $i \neq j$ and $(\alpha_i\alpha_j)^P = 1$ for some $p \geq 1$ then either

    (a) $(t_i t_j^{-1})^P = 1$

or

    (b) $t_1^{2k} = \cdots = t_n^{2k} = 1$ for some $k \geq 1$

Proof.   Since $(\alpha_i\alpha_j)^P = 1$ we have $(t_i t_j)^P \in N$. Since $\alpha_j^2 = 1$ we also have $(\alpha_i\alpha_j^{-1})^P = 1$ and therefore $(t_i t_j^{-1})^P \in N$. Then $(t_i t_j)^P = t_i^{2m} = t_j^{2m}$ for some m, and $(t_i t_j^{-1})^P = t_i^{2q} = t_j^{2q}$ for some q, because $N = \langle t_i^2 \rangle = \langle t_j^2 \rangle$.

Now $(t_i t_j^{-1})^P = t_i^{2q}$ gives $t_i^{-2p}(t_i t_j)^P = t_i^{2q}$, since the t_k^2 are central in Γ_P, and this together with $(t_i t_j)^P = t_i^{2m}$ gives $t_i^{2m} = t_i^{2(p+q)}$. If $m \neq p + q$ then $t_i^{2k} = 1$ for some $k \neq 0$, so $t_1^{2k} = \cdots = t_n^{2k}$. Otherwise $m = p + q$.

Then $(t_i t_j)^P = t_i^{2m}$ can be rewritten as

$$t_i(t_j(t_i t_j)^{P-1}) = t_i^{2m} = t_i t_i^{2m-1}$$

which gives

$$t_j(t_i t_j)^{P-1} = t_i^{2m-1} = t_i^{2m-2} t_i = t_j^{2m-2} t_i$$

and therefore

$$(t_i t_j)^{P-1} = t_j^{2m-3} t_i = t_j^{2m-4} t_j t_i = t_i^{2(m-2)} t_j t_i$$

If we iterate this procedure we get

$$t_i t_j = (t_i t_j)^{p-(p-1)} = t_i^{2(m-2(p-1))}(t_j t_i)^{p-1}$$

Now when we substitute this expression for $(t_j t_i)^{p-1}$ in $t_i^{-2p}(t_i t_j)^p = t_i^{2q}$ we get

$$t_i^{2q} = t_i^{-2p} t_i (t_j t_i)^{p-1} t_j = t_i^{-2p} t_i t_i^{-2(m-2(p-1))} t_i t_j^2 = t_i^{2(p-m)}$$

If $q \neq p - m$ we get $t_i^{2k} = 1$ for some $k \neq 0$ and so $t_1^{2k} = \cdots = t_n^{2k} = 1$. Otherwise $q = p - m$; this together with $m = p + q$ gives $q = 0$. That is, $(t_i t_j^{-1})^p = 1$.

The upshot of this lemma is that in looking for elements of Γ_p of the form $t_i t_j^{-1}$ with finite order we may as well look in $K_p = \Gamma_p / N$, a presentation for which is

$$K_p = \langle \alpha_1, \ldots, \alpha_n \mid \alpha_1^2 = \cdots = \alpha_n^2 = 1; \ \alpha_i \alpha_j = \alpha_k \alpha_\ell \text{ if } (i,j) \text{ and } (k,\ell) \text{ lie}$$
$$\text{in the same P-class} \rangle$$

We represent the presentation of K_p by an $n \times n$ array with entry $\alpha_i \alpha_j$ in the ij position when $i \neq j$, and 1 in the ii position, for $1 \leq i, j \leq n$, and we join the ij and $k\ell$ entries by a line if and only if (i,j) and (k,ℓ) lie in the same P-class.

EXAMPLE. (a) The simplest example of a group of the form K_p is

$$\langle \alpha_1, \alpha_2 \mid \alpha_1^2 = \alpha_2^2 = 1; \ \alpha_1 \alpha_2 = \alpha_2 \alpha_1 \rangle \cong \mathbb{Z}_2 \times \mathbb{Z}_2$$

This is one of a number of cases where K_p turns out to be a finite dihedral group. This group corresponds to the identifications given by the diagram of Figure 1.

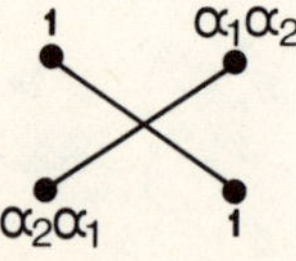

Figure 1

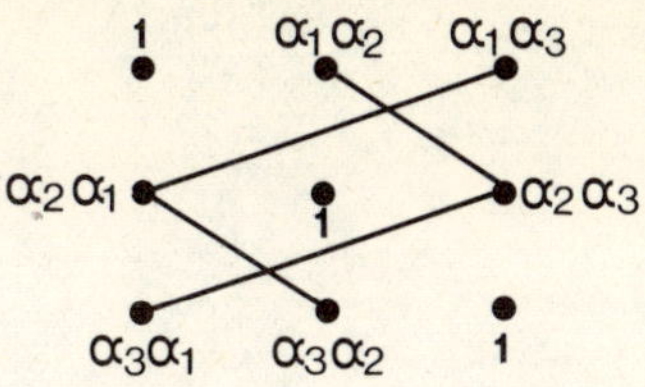

Figure 2

For n = 3, the diagram of Figure 2 gives us the group

$$\langle \alpha_1, \alpha_2, \alpha_3 \mid \alpha_1^2 = \alpha_2^2 = \alpha_3^2 = 1; \ \alpha_1\alpha_2 = \alpha_2\alpha_3 = \alpha_3\alpha_1 \rangle$$

(note that $\alpha_1\alpha_3 = \alpha_2\alpha_1 = \alpha_3\alpha_2$ is a consequence of the given relations) which is the dihedral group D_3 of order 6.

For n = 5 the diagram of Figure 3

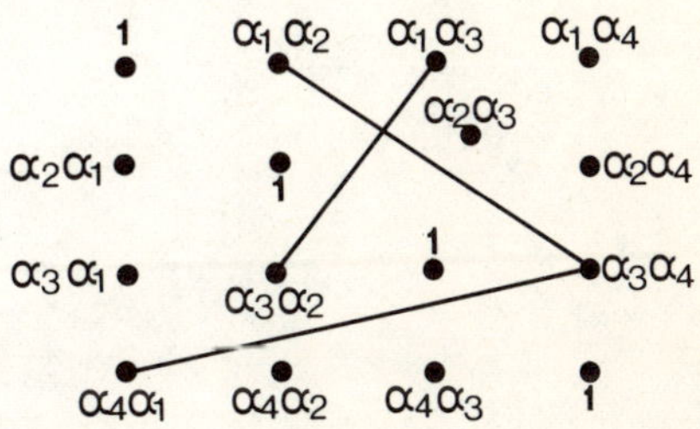

Figure 3

gives us the group

$$\langle \alpha_1, \alpha_2, \alpha_3, \alpha_4 \mid \alpha_1^2 = \alpha_2^2 = \alpha_3^2 = \alpha_4^2 = 1, \alpha_1\alpha_2 = \alpha_3\alpha_4 = \alpha_4\alpha_1, \ \alpha_1\alpha_3 = \alpha_3\alpha_2 \rangle$$

which is the dihedral group D_5 of order 10. This tells us that if among the relations of K_P we have (apart from the involutory relations)

$$\alpha_i\alpha_j = \alpha_k\alpha_\ell = \alpha_\ell\alpha_i, \ \alpha_i\alpha_k = \alpha_k\alpha_j$$

then $(\alpha_i\alpha_j)^5 = 1$ in K_P.

PROPOSITION. If the relations of K_P include

    (i) $\alpha_1\alpha_i = \alpha_{i-1}\alpha_{i+1}$ for $3 \leq i \leq n - 1$

   (ii) $\alpha_1\alpha_2 = \alpha_n\alpha_3$

 (iii) $\alpha_1\alpha_n = \alpha_{n-1}\alpha_2$

then K_P is finite.

    Proof.   The group K_P is a homomorphic image of

$$H = \langle \beta_1,\ldots,\beta_n \mid \beta_1^2 = \cdots = \beta_n^2 = 1; \ \beta_1\beta_i = \beta_{i-1}\beta_{i+1} \text{ for } 3 \leq i \leq n - 1;$$
$$\beta_1\beta_2 = \beta_n\beta_3; \ \beta_1\beta_n = \beta_{n-1}\beta_2 \rangle$$

We relabel $\beta_1 = t$, $\beta_1\beta_{i+1} = y_i$ for $1 \leq i \leq n - 1$ and write H as

$$\langle t,y_1,\ldots,y_{n-1} \mid t^2 = (ty_1)^2 = \cdots = (ty_{n-1})^2 = 1; \ y_1y_2 = y_3; \ldots;$$
$$y_{n-2}y_{n-1} = y_1; \ y_{n-1}y_1 = y_2 \rangle$$

The relations $(ty_i)^2 = 1$ can be rewritten as $ty_i = y_i^{-1}t$ and they give

$y_3^{-1}t = ty_3 = ty_1y_2 = y_1^{-1}y_2^{-1}t$ so $y_1y_2 = y_2y_1$. Since, as is easily seen, each

y_i belongs to the subgroup of H generated by y_1 and y_2 we have

$$H = \langle A(2,n-1),t \mid t^2 = (ty_1)^2 = \cdots = (ty_{n-1})^2 = 1 \rangle$$

where $A(2,n-1) = \langle y_1,\ldots,y_{n-1} \mid y_1y_2 = y_3; \ldots; y_{n-1}y_1 = y_2;$ and for $i < j$
$[y_i,y_j] = 1 \rangle$ is the $(n-1)$th abelianized Fibonacci group [2].

    By virtue of the relations $ty_i = y_i^{-1}t$ we can write each element of H
in the form $w(y_1,\ldots,y_{n-1})t^\varepsilon$ where ε is 0 or 1. Then $|H| \leq 2|A(2,n-1)|$ and
$|A(2,n-1)|$ is finite so H, and therefore K_P, is a finite group.

    EXAMPLE. The groups K_P are not all finite. An example of an infinite
group of this sort is given by the identifications of the diagram of Figure 4.

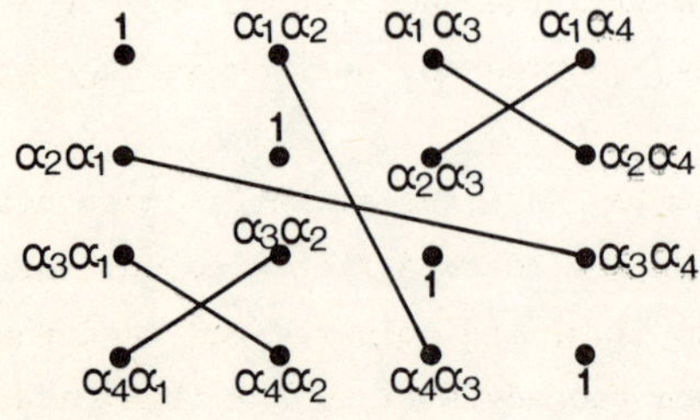

Figure 4

The group is
$$K_P = \langle \alpha_1, \alpha_2, \alpha_3, \alpha_4 \mid \alpha_i^2 = 1; \ \alpha_1\alpha_2 = \alpha_4\alpha_3; \ \alpha_1\alpha_3 = \alpha_2\alpha_4; \ \alpha_1\alpha_4 = \alpha_2\alpha_3 \rangle$$
$$= \langle \alpha_1, \alpha_2, \alpha_3 \mid \alpha_i^2 = 1; \ \alpha_1\alpha_2 = \alpha_2\alpha_1; \ (\alpha_1\alpha_2\alpha_3)^2 = 1 \rangle$$
$$= \langle \alpha_1, t, \alpha_3 \mid \alpha_1^2 = t^2 = \alpha_3^2 = 1; \ \alpha_1 t = t\alpha_1; \ \alpha_3 t = t\alpha_3 \rangle$$
$$\cong (\mathbb{Z}_2 * \mathbb{Z}_2) \times \mathbb{Z}_2$$

4. THE SUBGROUP GENERATED BY THE PRODUCTS OF GENERATORS

In the group K_P we expect that some $\alpha_i\alpha_j$ with $i \neq j$ will have finite order. We denote by Π_P the subgroup of K_P generated by the elements $\alpha_i\alpha_j$: specifically, we write $\xi_{ij} = \alpha_i\alpha_j$ and $\Pi_P = \langle \xi_{ij} : 1 \leq i, j \leq n \rangle$. The ξ_{ij} clearly satisfy the following relations:

(a) $\xi_{ii} = 1$ for all i

(b) $\xi_{ij}\xi_{jk} = \xi_{ik}$ for all i, j, k

(c) $\xi_{ij} = \xi_{k\ell}$ if (i,j) and (k,ℓ) are in the same P-class

(d) $\xi_{ik} = \xi_{j\ell}$ if (i,j) and (k,ℓ) are in the same P-class.

So Π_P is a homomorphic image of the group $\bar{\Pi}_P$ obtained by factoring the free group on n^2 generators ξ_{ij} ($1 \leq i, j \leq n$) by the normal subgroup corresponding to the relations (a) – (d). By virtue of these relations we can represent $\bar{\Pi}_P$ as the fundamental group of a 2-dimensional complex whose 1-skeleton is a *complete* graph.

Given a partition P of $\underline{n}^2$ as in section 2, for which all (i,i) comprise a single P-class, we define a coloured directed graph C_P as follows. As an undirected graph C_P is the complete graph on n vertices labelled $1,\ldots,n$: for each pair of distinct vertices i,j there is precisely one edge in C_P incident with both i and j, and there are no other edges. Each edge in C_P is directed and coloured so that the following conditions hold, where (i,j) stands for the edge in C_P with source i and target j:

(a) no two edges with the same source or target are coloured the same,

(b) each edge is coloured the same as an edge with a different source and target,

(c) if two triangles have two of their edges agreeing in colour and direction, then their third edges agree in colour and direction; precisely: if the edges (i,j) and (ℓ,m) are coloured red, and the edges (j,k) and (m,n) are coloured blue, then the edges (i,k) and (ℓ,n) are coloured the same.

(d) if two opposite sides of a square have the same colour then the remaining two edges have the same colour; precisely: if the edges (i,j) and (k,ℓ) with $i \neq \ell$ and $j \neq k$ are red then the edges (i,k) and (j,ℓ) have the same colour.

We now define a 2-complex by specifying the faces and their boundaries. Firstly each triangle is a face and the boundary of the triangle of Figure 5 is $[i,j]$, $[j,k]$, $[k,i]$ (where $[\alpha,\beta]$ stands for the colour of the edge whose vertices are α and β) whilst the boundary of the triangle of Figure 6 is $[i,j]$, $[j,k]$, $[k,i]^{-1}$.

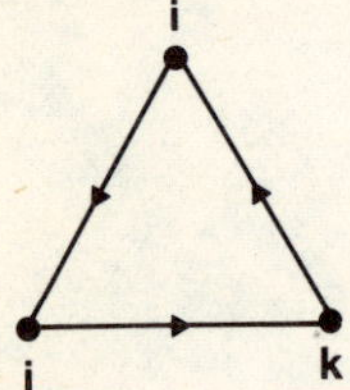

Figure 5

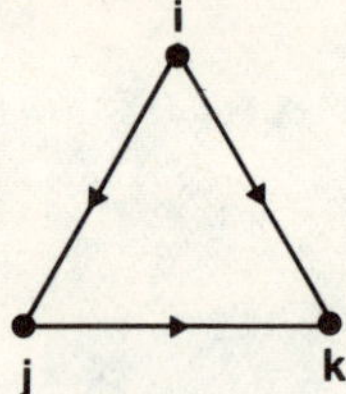

Figure 6

Let G be the group generated by the colours $[i,j]$, where $\overset{}{\underset{i}{\circ}}\!\longrightarrow\!\underset{j}{\circ}$ is a directed edge in C_p, subject to the relations $\partial = 1$ where ∂ is the boundary of a triangle in C_p. A sequence of Tietze transformations establishes that $G = \overline{\Pi}_p$.

Given a graph with coloured and directed edges, no loops and no multiple edges we perform substitution for a colour to get a new graph as follows:

SUBSTITUTION. If an edge from i to j is coloured blue and an undirected path π from i to j contains no blue edge then we erase the blue edge from i to j and replace all other occurrences of blue edges by a copy of the path π.

Substitution for blue in a coloured directed graph with specified faces F and boundaries provides us with a new coloured directed graph with faces as in Figure 7

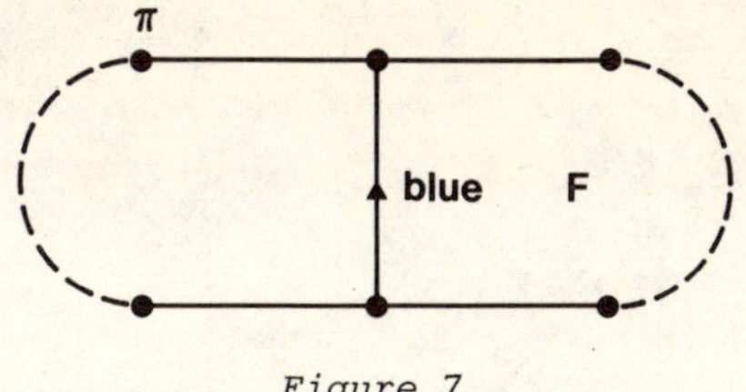

Figure 7

and obvious boundaries (namely the boundary sequence of F, minus blue, with π attached).

The group generated by the remaining colours with the boundaries of the new faces as relators is, by virtue of Tietze transformations, the same as the corresponding group for the original graph.

We therefore ask whether, for each position P, some finite sequence of substitutions for a colour starting from a C_P coloured according to (a) – (d) produces a graph with a closed directed path of a single colour of reduced length >1.

If an edge from i to j is coloured blue and there is an undirected path from i to j which is of reduced length m – 1 and contains no blue edge then we say that blue is eliminable by an m-gon.

Now we make some obvious remarks about the existence of uni-colour closed directed reduced paths in C_P as a result of a colouring satisfying (a), (b), (c) and (d).

Firstly we must have such a path if n – 1 edges of C_P are coloured blue. For the blue edges must form a directed path from vertex 1 to vertex n, say: if the edge from 1 to n, or from n to 1, were coloured blue then we would have a closed directed blue path of reduced length >1; if the edge from 1 to n were coloured red then some other edge would have to be coloured red and this would give us a closed directed blue path of reduced length >1.

We may assume therefore that C_P does not have n – 1 edges coloured blue.

Then we may also assume that every coloured edge of C_P is eliminable by triangles. Suppose to the contrary that blue is not eliminable by a triangle.

Then in C_p we have a triangle as in Figure 8 (a) or (b).

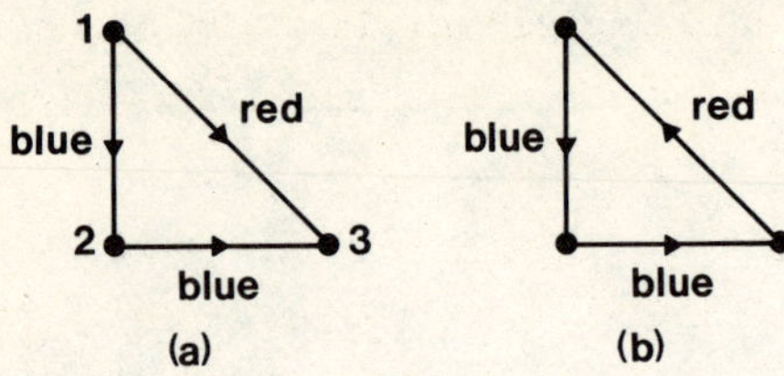

Figure 8

Assume case (a). Then C_p must contain a fourth vertex so consider the graph K_4 of Figure 9.

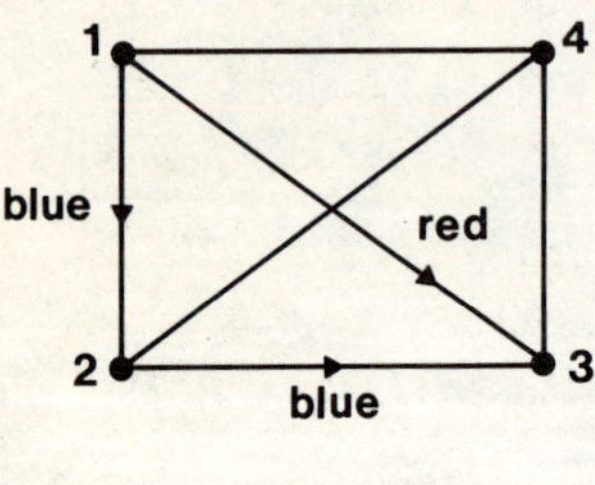

Figure 9

The edge from 4 to 2 cannot be coloured blue so colour it yellow. Since blue is not eliminable by a triangle the edge from 4 to 1 must be blue. Then yellow coincides with red so we have the graph of Figure 10.

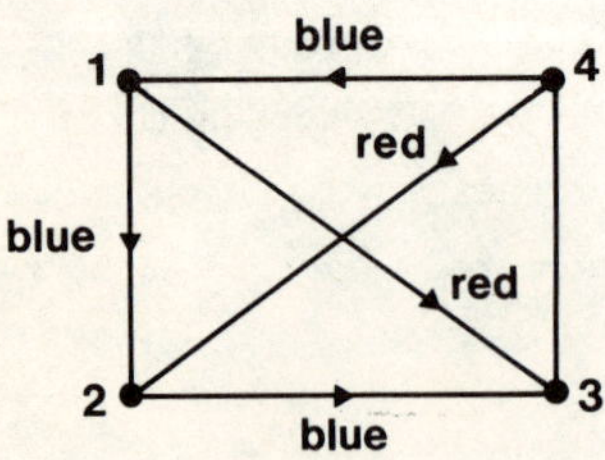

Figure 10

Now consider the edge from 3 to 4. If this is coloured orange (not blue) then blue is eliminable by a triangle so the edge must be coloured blue. This gives a closed directed blue path of reduced length 4.

Now assume case (b). If red coincides with blue then we have a closed directed blue path of reduced length 3. Otherwise we have a fourth vertex and then we proceed as in case (a).

Apart from these rather obvious remarks however we know almost nothing in general about the existence of closed uni-colour directed paths of reduced length >1 in the graphs obtained by substitutions from a C_p coloured according to (a) - (d).

REFERENCES

1. H. Bass, Traces and Euler characteristics, Homological Group Theory, Cambridge University Press, Cambridge, 1979, pp. 1-26.

2. D. L. Johnson, Presentation of groups, Cambridge University Press, Cambridge, 1976.

3. D. S. Passman, The algebraic structure of group rings, Wiley Interscience, New York, 1977.

SOLVABLE GROUPS AND THEIR CHARACTERS

I. MARTIN ISAACS*

Mathematics Department
University of Wisconsin
Madison, Wisconsin
U.S.A.

It is claimed by many optimists that the task of classifying all finite simple groups is now essentially complete; and it is feared by some pessimists
that if this is true, there is little of interest left to do in finite group
theory. Without taking a stand on the former opinion, it is my aim in this
lecture to give evidence that the fears of the pessimists are unfounded.

The only simple groups which are involved in solvable groups are those
of prime order, which, of course, have been known practically forever. Nevertheless, I will describe a class of problems about solvable groups, some of
which are only now beginning to become accessible. The problems I have in
mind involve characters.

I will digress to review enough of the basic facts so that the problems
to be discussed will, I hope, be intelligible to others than group theorists.

Let G be a finite group. Then Char(G) is a certain set of complex valued functions on G, each constant on conjugacy classes. (In fact, they are
the functions χ of the form

---

* At the time of writing the author was a Visiting Fellow at the Department
of Mathematics, I.A.S., Australian National University, Canberra, A.C.T.,
Australia.

$$\chi(g) = \text{trace}(X(g))$$

where X is a homomorphism of G into $GL(n, \mathbb{C})$, the group of nonsingular $n \times n$ complex matrices for various positive integers n.) For $\chi \in \text{Char}(G)$, the *degree* of χ is the value $\chi(1)$ at the group identity. It is always a positive integer. It follows that every character is a finite sum of *irreducible* characters, i.e. those which are not, themselves, sums of other characters. The set of irreducible characters of G is denoted $\text{Irr}(G)$.

It turns out that $\text{Irr}(G)$ is a basis for the space of those complex functions on G which are constant on conjugacy classes. In particular, $\text{Irr}(G)$ is finite and for each $\xi \in \text{Char}(G)$, the expression

$$\xi = \sum_{\chi \in \text{Irr}(G)} a_\chi \chi$$

is unique. Since sums of characters are characters, it follows that $\text{Char}(G)$ is exactly the set of nonnegative integer linear combinations of $\text{Irr}(G)$ with at least one nonzero coefficient. In fact, if for functions ϕ, $\theta: G \to \mathbb{C}$, one defines their "inner product"

$$[\phi, \theta] = \frac{1}{|G|} \sum_{g \in G} \phi(g)\overline{\theta(g)}$$

it turns out that $\text{Irr}(G)$ is an orthonormal set. It follows that a nonzero function $\xi: G \to \mathbb{C}$ which is constant on conjugacy classes, is a character if and only if $a_\chi = [\xi, \chi]$ is a nonnegative integer for each $\chi \in \text{Irr}(G)$.

Because of the above remarks, we focus attention on $\text{Irr}(G)$. The (necessarily irreducible) characters of degree 1, the *linear* characters, are especially easy to understand: they are exactly the homomorphisms $G \to \mathbb{C}^\times$. A natural question, then, is which groups have all irreducible characters linear. The answer is that $\chi(1) = 1$ for all $\chi \in \text{Irr}(G)$ if and only if G is abelian.

How can we understand the nonlinear irreducible characters of a group? Some of these may be "induced" from proper subgroups and so we need to consider the process of character induction. Let $H \subseteq G$ and let $\phi: H \to \mathbb{C}$ be a function. Define $\phi^O: G \to \mathbb{C}$ by

$$\phi^O(g) = \begin{cases} \phi(g) & \text{if } g \in H \\ 0 & \text{if } g \notin H \end{cases}$$

and write

$$\phi^G(g) = \frac{1}{|H|} \sum_{x \in G} \phi^O(xgx^{-1})$$

Then ϕ^G is the *induced* function $G \to \mathbb{C}$. It is trivial to check the "Frobenius reciprocity" formula

$$[\phi^G, \Theta] = [\phi, \Theta_H]$$

for functions $\phi: H \to \mathbb{C}$ and $\Theta: G \to \mathbb{C}$, each constant on conjugacy classes, where Θ_H denotes the restriction of Θ to H.

It is easy to see that if $\xi \in$ Char(G), then $\xi_H \in$ Char(H), and it follows by Frobenius reciprocity that if $\Psi \in$ Char(H) then $\Psi^G \in$ Char(G). The proof is as follows.

Proof.   Observe that $\Psi^G(1) = |G|\Psi(1)/|H| \neq 0$ and that Ψ^G is constant on conjugacy classes. Therefore, all we need to show is that $[\Psi^G, \chi]$ is a nonnegative integer for all $\chi \in$ Irr(G). However, $[\Psi^G, \chi] = [\Psi, \chi_H]$ which, being the inner product of two characters of H, is a nonnegative integer as required.

Of course, even if $\Psi \in$ Irr(H), it need not be true that $\Psi^G \in$ Irr(G). Nevertheless, given $\chi \in$ Irr(G), there may exist a proper subgroup $H < G$ such that $\chi = \Psi^G$ for some $\Psi \in$ Char(H). (Necessarily, $\Psi \in$ Irr(H).) Such irreducible characters χ are said to be *imprimitive* and the others, not induced from any proper subgroup, are, of course, the *primitive* characters. Finally, $\chi \in$ Irr(G) is said to be *monomial* if $\chi = \lambda^G$ where λ is a linear character of some (not necessarily proper) subgroup of G. Thus the monomial characters are all of those irreducible characters which are either linear or are induced from linear characters of proper subgroups.

Since we understand linear characters, and we know all the groups for which all irreducible characters are linear, it seems reasonable to ask which groups have the property that all of their irreducible characters are monomial. These groups are called M-*groups* and it would be desirable to be able to give a purely group-theoretic description of this class of groups. In other words, we want a condition which does not mention characters, and which is necessary and sufficient for a group to be an M-group. At the present state of our knowledge, this appears to be hopeless.

To bring the topic of discussion back to the title of this lecture, we mention the fundamental result of Taketa [9], proved in 1930: All M-groups are solvable. This suggests a comparison of the class of all M-groups with other, more well known classes of finite groups. We have the following chain of proper containments:

$$\left\{\begin{array}{l}\text{Abelian}\\\text{groups}\end{array}\right\} < \left\{\begin{array}{l}\text{Nilpotent}\\\text{groups}\end{array}\right\} < \left\{\begin{array}{l}\text{Supersolvable}\\\text{groups}\end{array}\right\} < \left\{\begin{array}{l}\text{M-groups}\end{array}\right\} < \left\{\begin{array}{l}\text{Solvable}\\\text{groups}\end{array}\right\}$$

The last inequality is Taketa's theorem and the fact that it is proper is demonstrated by the group SL(2, 3) of order 24 which is not an M-group. That all supersolvable groups are M-groups follows from a more general sufficient condition proved by Huppert [6]. Another group of order 24, the symmetric group S_4, shows that this containment is proper also, since S_4 satisfies Huppert's condition and so is an M-group, and yet is not supersolvable.

That Huppert's condition is very far from necessary for a group to be an M-group is seen by the fact that subgroups of groups of Huppert type are themselves of this type, while subgroups of M-groups need not be M-groups. In fact, it is a theorem that one can make no statement about which groups can occur as subgroups of M-groups beyond that they must be solvable. This is because given any solvable group, Dade showed how to construct an M-group containing it (see page 585 of [7]).

It appears that the reason that M-groups are so difficult to characterise is that their subgroups are not under control. For instance, although more general sufficient conditions than Huppert's are known, no reasonable sufficient condition has been found which is not inherited by subgroups.

Giving up (at least temporarily) on trying to characterize M-groups, let us focus on what appears to be the source of the difficulty, namely the subgroups of M-groups. Despite Dade's construction, there are things which can be said about particular types of subgroups. Dornhoff [4] and Seitz [9] (independently) showed that if G is an M-group and $H \subseteq G$ is such that

(a) H is normal and

(b) $(|H|, |G:H|) = 1$

then H is an M-group. This immediately suggests the question of whether or not either condition (a) or condition (b) by itself is sufficient to guarantee that H is an M-group.

Nothing seems to be known about the sufficiency of (b) and so I will devote the remainder of this lecture to the consideration of normal subgroups of M-groups.

Dade [3] constructed an example of an M-group of order $2^9 \cdot 7$ which has a normal subgroup of index 2 that is not an M-group. The involvement of the

prime 2 in this group is not accidental. There does not exist a group of
odd order constructed analogously, and this leaves open the problem of whether or not normal subgroups of M-groups of odd order must be M-groups.
Slightly more ambitiously, one might conjecture that if G is an M-group and
$H \lhd G$ is not an M-group then, *both* $|H|$ and $|G : H|$ are even.

It was claimed by Čubarov [1] that he has succeeded in proving this con-
jecture. In fact, he claims more, namely if $\chi \in \mathrm{Irr}(G)$ is monomial and $H \lhd G$,
then under suitable oddness and solvability hypotheses, and without assuming
that G is an M-group, one has that all of the irreducible characters of H involved in χ_H are monomial. If this were true, it would certainly prove the
conjecture. Unfortunately, this stronger assertion is simply false, as is
demonstrated by an example shown to me by Berger. It is my opinion that
despite Čubarov's work, the problem must still be considered to be open.

I have recently obtained a result (not yet published) which gives some
evidence in favor of the conjecture and I will close by describing that result. Observe that the only primitive characters which an M-group can have
are the linear characters, and so it is reasonable to consider what primitive
characters a normal subgroup of an M-group can have.

THEOREM.   Let G be an M-group and let $H \lhd G$.   Suppose $\Theta \in \mathrm{Irr}(H)$ is primitive.
Then

    (a) $\Theta(1)$ is a power of 2.

    (b)   If $\Theta(1) > 1$, then $2 \mid |G : H|$.

Since for any group, the degree of an irreducible character must divide
the order of the group, it follows in the situation of the theorem that if
either $|H|$ is odd or $|G : H|$ is odd, then $\Theta(1) = 1$, whenever Θ is a primitive
character of H. This is what one would expect if the conjecture is true.

The proof of this theorem uses a great deal of the rapidly growing theory
of characters of solvable groups, and lies quite deep. Among other things,
it uses the ideas of Gajendragadkar [5] and also parts of the theory of character correspondences developed by Dade [2] and myself [8].

I hope that I have succeeded in making the point that even in groups
where we know all of the relevant simple groups, there exist interesting,
difficult problems, and much work remains to be done.

REFERENCES

1. I. A. Čubarov, Monomial characters and subgroups, Russian Math. Surveys
    33 (1978), 143-144.

2. E. C. Dade, Characters of solvable groups, mimeographed notes, Univ. of
    Illinois, Urbana, 1967.

3. E. C. Dade, Normal subgroups of M-groups need not be M-groups, Math.
    Zeit. 133 (1973), 313-317.

4. L. Dornhoff, M-groups and 2-groups, Math. Zeit. 100 (1967), 226-256.

5. D. Gajendragadkar, A characteristic class of characters of π-separable
    groups, J. of Algebra 59 (1979), 237-259.

6. B. Huppert, Monomiale Darstellungen endlicher Gruppen, Nagoya Math. J.
    6 (1953), 93-94.

7. B. Huppert, Endliche Gruppen I, Springer-Verlag, Berlin, 1967.

8. I. M. Isaacs, Characters of solvable and symplectic groups, Amer. J. of
    Math. 95 (1973), 594-635.

9. G. Seitz, M-groups and the supersolvable residual, Math. Zeit. 110
    (1969), 101-122.

10. K. Taketa, Uber die Gruppen deren Darstellungen sich sämtlich auf monom-
    iale Gestalt transformieren lassen, Proc. Jap. Imp. Acad. 16 (1930),
    31-33.

GRAPHS AND THEIR AUTOMORPHISM GROUPS

CHERYL E. PRAEGER

Department of Mathematics
University of Western Australia
Nedlands, Western Australia
Australia

This paper is concerned with finite simple connected undirected graphs Γ with a group G of automorphisms such that

    (a)   G is transitive on the set $V\Gamma$ of vertices of Γ,
and

    (b)   for $a \in V\Gamma$, the stabilizer G_a is 2-transitive on the set $\Gamma_1(a)$ of vertices joined to a.

(See H. Wielandt's book [11] for notation.)

Such graphs have been studied recently by several people, for example P. J. Cameron [2] and A. Gardiner [6]. Clearly a graph Γ with this property is regular of valency v and to eliminate trivial cases let us assume that $v \geq 3$. Also if Γ contains a triangle then Γ is a complete graph so we shall assume that Γ contains no triangles, that is, the length of the shortest circuit in Γ is at least 4. Then if we call the length of the shortest path between two vertices a and b the *distance* $d(a,b)$ between a and b, any pair of vertices at distance 2 has the same number k of paths of length 2 between them, (see [8] Lemma 2). This parameter k gives a measure of the number of quadrilaterals, (that is, circuits of length 4) in Γ, and k = 1 if and only if Γ contains no quadrilaterals. The problem we are concerned with here is:

PROBLEM. Find all graphs Γ with k "large".

Exactly how "large" k should be before we can classify Γ is a difficult question, but Peter Cameron has conjectured:

CONJECTURE.   (P. J. Cameron; private communication)   If G is primitive on $V\Gamma$
then $k \leq 6$.

(There is a graph with $k = 6$ on 100 vertices satisfying our hypotheses; its
automorphism group is the Higman-Sims simple group (see [1] Proposition
3.4).)   To understand the status of the problem let us consider the known
partial solutions.

(1)   (P. J. Cameron [2], Theorems 4.1, 4.2.)   If $k > v/2$ then Γ is the incidence graph of a (possibly degenerate) self-dual symmetric design D satisfying

     (a)   Aut D is 2-transitive on the points of D;
and
     (b)   if b is a block of D then $(\text{Aut } D)_b$ is 2-transitive on the points
incident with b.

    The next three results show that extra restrictions on G_a give much
better results.

(2)   (P. J. Cameron [2], Theorem 4.4.)   If G_a is 3-transitive on $\Gamma_1(a)$ then
either

     (a) Γ is known,
or
     (b) $k \leq 2$,
or
     (c) $v = (x + 1)(x^2 + 5x + 5)$, $k = (x + 1)(x + 2)$ for some positive
integer x and Γ is strongly regular on $(x + 1)^2(x + 4)^2$ vertices or is
obtained from such a graph by doubling.

(3)   (P. J. Cameron [2], Theorem 4.5.)   If G_a is 4-transitive on $\Gamma_1(a)$ then
either $k \leq 2$ or Γ is known.   Moreover if G_a acts on $\Gamma_1(a)$ as A_v or S_v then
either $k = 1$ or Γ is known.

(4)   (C. E. Praeger [8], and P. J. Cameron and C. E. Praeger [4].)   If
$$\text{PSL}(n,q) \leq G_a^{\Gamma_1(a)} \leq \text{P}\Gamma\text{L}(n,q)$$ for some $n \geq 3$ and prime power q in its natural
representation on the $v = (q^n - 1)/(q - 1)$ points of the projective geometry,
then either

(a) Γ is known

or

(b)   k is 1,2,3,4 or 6. Moreover if k > 1 then G_a acts faithfully on $\Gamma_1(a)$.

To get comparable results when $G_a^{\Gamma_1(a)}$ is an arbitrary 2-transitive group seems to be very difficult. However it is widely believed that the finite simple groups will soon be classified and this classification will allow the classification of all finite 2-transitive permutation groups, (see Cameron [3], Theorem 5.3(S)). In fact it is possible that we know all of the 2-transitive groups already. Thus one way of tackling the problem is to examine the situation for each of the known 2-transitive groups*. This has already been done in the case where G_a is not faithful on $\Gamma_1(a)$ with satisfactory results.

(5)   (See P. J. Cameron and C. E. Praeger [4], Corollary 2.) If $G_a^{\Gamma_1(a)}$ is not faithful and is one of the known 2-transitive groups then either Γ is known, or k = 1, or $G_a^{\Gamma_1(a)}$ has a regular normal subgroup.

(For a list of the known 2-transitive groups see Praeger [7].) The case where $G_a^{\Gamma_1(a)}$ has a regular normal subgroup seems difficult to handle at present. Apart from this case we are left with the situation where G_a is faithful on $\Gamma_1(a)$. The results (2) and (3) above of Cameron effectively deal with all the known 3-transitive groups except where $PSL(2,q) \leq G_a^{\Gamma_1(a)} \leq P\Gamma L(2,q)$ in its usual representation, q a prime power, (or the case of a regular normal subgroup). Result (4) deals with the higher dimensional linear groups, and Cameron has examined the only other known 2-transitive subgroup of $P\Gamma L(n,q)$, $n \geq 3$, namely $A_7 \leq P\Gamma L(4,2)$:

(6)   (P. J. Cameron; private communication.) If $G_a^{\Gamma_1(a)} \simeq A_7$ in its representation on the 15 points of the projective space $PG(3,2)$ then either k ≤ 3, or k = 5, G = $P\Sigma U(3,5)$, and Γ is a bipartite graph on 100 vertices.

---

* Since this paper was submitted P. J. Cameron and the author have determined the graphs with k > 6 associated with all the known 2-transitive groups which do not have a regular normal subgroup.

These are all the results to date. We shall deal here with the case of the one-dimensional projective linear groups.

(7)   If $PSL(2,q) \leq G_a^{\Gamma_1(a)} \leq P\Gamma L(2,q)$, q a prime power, in its usual representation of degree q + 1, then either Γ is known or $k \leq 2$.

More precisely we prove the following:

THEOREM. Let Γ be a finite simple connected incomplete undirected graph with a group G of automorphisms which is transitive on vertices and such that, for $a \in V\Gamma$, $PSL(2,q) \leq G_a^{\Gamma_1(a)} \leq P\Gamma L(2,q)$ for some prime power q, in its usual representation of degree q + 1. Then one of the following is true, (where k is the number of paths of length 2 between two vertices at distance 2).

(a) $k \leq 2$, and if $k = 2$, G_a is faithful on $\Gamma_1(a)$.

(b) Γ is $K_{q+1,q+1}$, and either $PSL(2,q)$ wr $Z_2 \leq G \leq P\Gamma L(2,q)$ wr Z_2, or G_a is faithful on $\Gamma_1(a)$ and in this case q is 2,3,5 or 7.

(c) Γ is $K_{q+2,q+2}$ with the edges of a matching removed, and q is 2,3,4 or 9.

(d) $k = 6$, $q = 9$, G_a is $PSL(2,9) \simeq A_6$, or S_6 and G is an extension of an elementary abelian group of order 32 by G_a. If Δ is the graph obtained from the cube Q_6 by identifying antipodal vertices, then Γ is the graph with $V\Gamma = V\Delta$ and two vertices a,b are joined in Γ if and only if they are at distance 3 in Δ.

(e) $k = 4$, $q = 7$, $G_a \simeq PSL(2,7)$, and either Γ is the incidence graph of the complementary design of points and hyperplanes of PG(3,2), and $G \cong S_7$, or Γ is a double cover of this graph and $G \cong S_7 \times Z_2$.

I am grateful to Drs. C.A. Rowley and L. Kovacs and Professor M. Isaacs for their help in the proof of this theorem. Chris Rowley constructed a group of automorphisms of the graph $K_{8,8}$ with stabilizer isomorphic to $PSL(2,7)$. This construction is given in the proof of the theorem. It is interesting because the analogous question of the existence of a group of automorphisms of the graph obtained from $K_{8,8}$ by doubling, with stabilizer $PSL(2,7)$, was answered negatively by Laci Kovacs and Martin Isaacs. The proof of this given later is due to Laci Kovacs.

If we do not assume that Γ is undirected we can formulate our problem as a problem about primitive permutation groups. The analogous result is:

COROLLARY. Let G be a primitive permutation group on a finite set Ω such that for a $\in \Omega$, G_a has an orbit $\Gamma(a)$ with $PSL(2,q) \leq G_a^{\Gamma(a)} \leq P\Gamma L(2,q)$ in its natural representation on $|\Gamma(a)| = q + 1$ points for some prime power q. Then $\Gamma \circ \Gamma^*(a) = \{b; \Gamma(a) \cap \Gamma(b)$ is nonempty$\}$ is an orbit of G_a of length $q(q + 1)$ or $q(q + 1)/2$, (and in the latter case G_a is faithful on $\Gamma(a)$).

Proof of the Theorem.   Let G, Γ satisfy the hypotheses of the theorem and let X be the normal subgroup of G_a isomorphic to $PSL(2,q)$. By [2], $|\Gamma_2(a)| = (q + 1)q/k$, (where $\Gamma_i(a)$ denotes the set of points at distance i from a). If G_a is unfaithful on $\Gamma_1(a)$ then by Cameron and Praeger [4] k = 1 or (b) is true. So assume that G_a is faithful on $\Gamma_1(a)$. If $b \in \Gamma_1(a)$ then G_{ab} is transitive of degree q on the two sets $\Gamma_1(a) - \{b\}$ and $\Gamma_1(b) - \{a\}$ with a regular normal subgroup. By [11], 11.2 these two representations are equivalent and so if $c \in \Gamma_1(b) - \{a\}$ then G_{abc} fixes a point b' of $\Gamma_1(a) - \{b\}$; so $G_{abc} = G_{abb'}$ has orbit lengths $1,1,q - 1$ or $1,1,(q - 1)/2,(q - 1)/2$ in $\Gamma_1(a)$. Now since G_{ab} is transitive on $\Gamma_1(b) - \{a\} = \Gamma_1(b) \cap \Gamma_2(a)$ of degree q it follows that G_{ac} is transitive on $\Gamma_1(c) \cap \Gamma_1(a)$ of degree k. The latter set is the orbit of G_{ab} containing b and is a union of some of the orbits of G_{abc} including $\{b\}$. Hence k is $1,2,(q + 1)/2,(q + 3)/2,q$ or $q + 1$. Let us suppose that $k \geq 3$.

Case k = q + 1

Here Γ is the complete bipartite graph $K_{q+1,q+1}$ and X = $PSL(2,q)$ is transitive on $\Gamma_2(a)$ of degree q. It follows from Dickson [5], p.286-87, that q is 2,3,5,7 or 11. Further G has a normal subgroup Y of index 2 which is 2-transitive on $\Gamma_1(a)$ and $\{a\} \cup \Gamma_2(a)$. Since $G_a = Y_a$ is faithful on $\Gamma_1(a)$, either the kernel of the action of Y on $\Gamma_1(a)$ is trivial or it acts regularly on $\{a\} \cup \Gamma_2(a)$. In the latter case q is 2 or 3 and it is easy to construct an automorphism group G of Γ with these properties. In the former case q is 5,7 or 11, Y is a transitive extension of the representation of $Y_a = G_a$ of degree q, and Y has two conjugacy classes of subgroups isomorphic to $G_a = Y_a$ which are all conjugate to G. This is possible if q = 5, for example if G = Aut S_6, Y = S_6, and is impossible if q = 11, since Y is M_{11} which has no suitable automorphism group G. Finally if q = 7, then Y is the affine group AGL(3,2), and N = $C_G(O_2(Y))$ is a normal abelian subgroup of G of order

16.   The vertices of Γ are in 1-1 correspondence with the elements of N and the actions of $X = G_a$ on $V\Gamma$ and by conjugation on N are equivalent (see [11], 11.2).   Thus X has orbits on N of lengths 1,7,8, and $0_2(Y)$ is the only proper nontrivial X-invariant subgroup, $|0_2(Y)| = 8$.

We show by a construction due to Chris Rowley that such a group of automorphisms of $K_{8,8}$ exists: Let N be an elementary abelian group of order 16 with a subgroup V of order 8. The aim is to extend the natural linear action of $X \simeq GL(3,2)$ on V to a linear action on N in such a way that $X \simeq PSL(2,7)$ acts on $N\backslash V$ as on the projective line L over $GF(7)$. Once this is done we can define Γ to be the graph with vertex set N and edges $\{n,un\}$ where $n \in N$, $u \in N\backslash V$. Clearly Γ is $K_{8,8}$ and $G = N.X$ is an automorphism group of Γ satisfying our hypotheses. Defining the action of X on N exploits the fact that $GL(3,2)$ and $PSL(2,7)$ are isomorphic, and an explicit isomorphism which we shall use is given by:

LEMMA.   The group X has a presentation

$$X = \langle x,y,z \,|\, x^7 = y^3 = z^2 = 1, \; y^{-1}xy = x^2, \; z^{-1}yz = y^2 \rangle$$

(a)   The map $\phi : X \to GL(3,2)$ determined by

$$\phi(x) = \begin{pmatrix} 0 & 1 & 0 \\ 0 & 0 & 1 \\ 1 & 1 & 0 \end{pmatrix}, \quad \phi(y) = \begin{pmatrix} 1 & 0 & 0 \\ 0 & 0 & 1 \\ 0 & 1 & 1 \end{pmatrix}, \quad \phi(z) = \begin{pmatrix} 1 & 0 & 0 \\ 0 & 0 & 1 \\ 0 & 1 & 0 \end{pmatrix}$$

is an isomorphism.

(b)   The map $\psi : X \to PSL(2,7)$ determined by $\psi(x) = (\underline{0}\ \underline{1}\ \underline{2}\ \underline{3}\ \underline{4}\ \underline{5}\ \underline{6})$, $\psi(y) = (\underline{1}\ \underline{2}\ \underline{4})(\underline{3}\ \underline{6}\ \underline{5})$, $\psi(z) = (\underline{0}\ \infty)(\underline{1}\ \underline{6})(\underline{2}\ \underline{3})(\underline{4}\ \underline{5})$, where $L = \{\infty, \underline{0}, \underline{1}, \ldots, \underline{6}\}$ is the projective line over $GF(7)$, is an isomorphism.

Proof.   Straightforward verification.

Now X acts linearly on V by identifying V with the set of ordered triples of elements of $GF(2)$ and identifying X with $\phi(X)$. Let us denote the image of $v \in V$ under $g \in X$ in this action by v^g. Next define a 1-1 correspondence f between the projective line L and V by $f(\infty) = (0,0,0)$ and for $0 \le n \le 6$, $f(\underline{n}) = (1,0,0)^{x^n}$, with x as in the presentation for X in the lemma. If we identify the corresponding points of V and L and identify X with $\psi(X)$ then X also acts on V as $PSL(2,7)$ and we shall denote the image of $v \in V$ under $g \in X$ in this new action by vg.

LEMMA. For all $v \in V$ and $g \in G$, $vg = v^g + \infty g$.

Proof. This is true for all v if $g = x$ (by the definition of f) or if $g = z$ (by an easy verification). Now suppose that for all $v \in V$, $vg = v^g + \infty g$ and $vh = v^h + \infty h$ for some $g, h \in X$. If this implies that for all $v \in V$, $vgh = v^{gh} + \infty gh$, then, since X is generated by x and z, the result follows. Now $vgh = (vg)h = (v^g + \infty g)h = (v^g + \infty g)^h + \infty h = v^{gh} + (\infty g)^h + \infty h$ (since h is acting linearly on V) $= v^{gh} + \infty gh$. This completes the proof.

Now, to extend the linear action of X on V to a linear action of X on N, let u be a fixed element of $N \backslash V$. For $g \in X$ and $v \in V$ define $(v + u)^g = v^g + \infty g + u$. Each element g of X is clearly an automorphism of N. To check that we have defined a linear action it is sufficient to show that for all $v \in V$ and $g, h \in X$, $(v + u)^{gh} = ((v + u)^g)^h$. This is true since $((v + u)^g)^h = (v^g + \infty g + u)^h = (v^g + \infty g)^h + \infty h + u = v^{gh} + (\infty g)^h + \infty h + u = v^{gh} + \infty gh + u$ (by the lemma) $= (v + u)^{gh}$.

Thus $K_{8,8}$ has a group of automorphisms with the required properties.

Case $k = q \geq 3$

Here $|\Gamma_1(a)| = |\Gamma_2(a)| = q + 1$ and the full subgraph of Γ with vertex set $\Gamma_1(a) \cup \Gamma_2(a)$ is the complete bipartite graph $K_{q+1,q+1}$ minus a matching. Furthermore the actions of G_a on $\Gamma_1(a)$ and $\Gamma_2(a)$ are equivalent. Thus G_a is 2-transitive on $\Gamma_2(a)$ and it follows that $|\Gamma_3(a)| = 1$ and Γ is $K_{q+2,q+2}$ minus a matching. As in the previous case G has a subgroup Y of index 2 which acts on $\{a\} \cup \Gamma_2(a)$ as a transitive extension of the representation of $Y_a = G_a$ of degree $q + 1$. Hence by [10] q is 2,3,4 or 9 and in each case $G \cong Y \times Z_2$ is an automorphism group of Γ with the required properties.

Case $k = (q + 3)/2$

Here q must be odd. Since k divides $(q + 1)q$ it follows that q is 3 or 9. If $q = 3$ then $k = q$ and this case was dealt with above. So assume that $q = 9$. Then $|\Gamma_2(a)| = 15$ and as G_a has no transitive representation of degree 3 or 5, G_a is primitive on $\Gamma_2(a)$. By Sims [9] G_a is isomorphic to $A_6 \simeq PSL(2,9)$ or S_6, and it follows that G_{ac} has orbits of length 1,6,8 in $\Gamma_2(a)$, and hence $\Gamma_1(c) - \Gamma_1(a)$ is a G_{ac}-orbit of length 4 in $\Gamma_3(a)$. Thus G_a is transitive on $\Gamma_3(a)$ and $|\Gamma_3(a)| = 60/j$, where $j = |\Gamma_2(a) \cap \Gamma_1(d)|$ for $d \in \Gamma_3(a)$. Also $6 = k \leq j \leq 10$ so j is 6 or 10. If $j = 6$ then G_{ad} has orbits of length 1,9 in $\Gamma_1(a)$ and hence in $\Gamma_1(d)$; this is impossible as $\Gamma_2(a) \cap \Gamma_1(d)$ is an orbit of G_{ad} of length j. Hence j = 10, $|V\Gamma| = 32$ and G_a acts as A_6 or S_6 on $\Gamma_3(a)$ or degree 6. Part (d) follows from Cameron [2], Theorem 4.5.

Case k = (q + 1)/2

Here q must be odd and as k $\geq$ 3, q $\geq$ 5. Since X_b is transitive on $\Gamma_1(b)$ - {a}, X = PSL(2,q) is transitive on $\Gamma_2(a)$ of degree 2q and so by Dickson [5], p.285, either q = 5 and X is primitive on $\Gamma_2(a)$, or q = 7 and X has 7 blocks of imprimitivity of length 2 in $\Gamma_2(a)$.

Let q = 5. Then G_a is A_5 or S_5, but as G_{ac} has two orbits of length 3 in $\Gamma_1(a)$ we must have $G_a = A_5$. If Γ had diameter 2 then G would be primitive on $V\Gamma$ of prime degree 17, but neither 2-transitive nor metacyclic, contradicting Burnside's Theorem (see [11], 11.7). Hence the diameter is at least 3 and $|\Gamma_3(a)| = 30/j$ where j = $|\Gamma_2(a) \cap \Gamma_1(e)| \geq 3$ for e $\in \Gamma_3(a)$. So j is 3,5 or 6, and $|\Gamma_3(a)|$ is 10,6 or 5; also in the first case either Γ has diameter 3 or G_a is transitive on $\Gamma_4(a)$. Suppose that $\Gamma_2(a),\ldots,\Gamma_i(a)$ are all G_a-orbits of length 10, where i $\geq$ 2. Then by similar arguments G_{ae}, e $\in \Gamma_i(a)$, has two orbits of length 3 in $\Gamma_1(e)$ so that either Γ has diameter i or $\Gamma_{i+1}(a)$ is a G_a-orbit of length 10,6, or 5 as $|\Gamma_i(a) \cap \Gamma_1(f)|$ is 3,5 or 6 respectively, where f $\in \Gamma_{i+1}(a)$. Now let $\Gamma_j(a)$, $2 \leq j \leq t + 1$, be G_a-orbits of length 10, and $\Gamma_{t+2}(a)$ be empty or contain less than 10 vertices, where t $\geq$ 1. Let d be the diameter. Then either (a) d = t + 1 $\geq$ 3, or (b) d = t + 2, $|\Gamma_{t+2}(a)| = 5$, or (c) d = t + 3, $|\Gamma_{t+2}(a)| = 6$, $|\Gamma_{t+3}(a)| = 1$. A Sylow 3-subgroup P of G_a has f = t + 1,t + 3, fixed points and 3t + 2,3t + 3, orbits of length 3 in the cases (a),(b) respectively. By Witt's Lemma [11], 3.7, $|N_G(P)| = f|N_{G_a}(P)| = 6f$, and this divides $|G| = 60|V\Gamma|$. If 5 divides f then there is a 5-element which centralises P and must fix point-

wise at least 4 P-orbits of length 3. This is impossible as 5-elements of G fix at most 2 points. Thus f divides $2|V\Gamma|$. In case (a) it follows that t is 2 or 5. These two values are shown to be impossible by a similar argument using a subgroup of G_a of order 2. In case (b), we find that t + 3 divides 36. Similar considerations for a subgroup of G_a or order 2 show that 2t + 5 divides 3.5.13, and no values of t satisfy both conditions. In case (c) Γ is antipodal and if t > 1 its derived graph is of type (a) or (b) which is impossible. Hence t = 1. However in this case {a} $\cup \Gamma_4(a)$ is a block of imprimitivity for G of length 2, and $\Gamma_2(a)$ is a union of five blocks of length 2 which contradicts the fact that G_a is primitive on $\Gamma_2(a)$.

Thus $q = 7$. Here X is transitive on $\Gamma_2(a)$ of degree 14, and as PGL(2,7) has no subgroup of index 14 which is not contained in PSL(2,7), it follows that $G_a = X = PSL(2,7)$. Then $G_{ac} \simeq A_4$ has two orbits of length 4 in $\Gamma_1(c)$ and no orbit of length 4 in $\Gamma_2(a)$, so $\Gamma_1(c) - \Gamma_1(a) \subseteq \Gamma_3(a)$ and $|\Gamma_3(a)| = 56/j$ where $j = |\Gamma_2(a) \cap \Gamma_1(e)|$ for $e \in \Gamma_3(a)$. Also $4 = k \leq j \leq 8$ and so j is 4,7 or 8 and in the first case it follows as above that the diameter $d > 3$ and G_a is transitive on $\Gamma_4(a)$. Using similar arguments to the previous paragraph we may assume that $\Gamma_2(a),...,\Gamma_{t+1}(a)$ are all G_a-orbits of length 14 and that $\Gamma_{t+2}(a)$ is a G_a-orbit of shorter length, where $t \geq 1$. Then either (a) $d = t + 2$, $|\Gamma_{t+2}(a)| = 7$, or (b) $d = t + 3$, $|\Gamma_{t+2}(a)| = 8$, $|\Gamma_{t+3}(a)| = 1$. A Sylow 3-subgroup of G_a fixes $f = 2t + 4, 2t + 6$ vertices respectively and similar arguments to those for $q = 5$ above show that f divides $28|V\Gamma|$ and that 7 does not divide f. In case (a) this shows that t is 1,2,4,6,10 or 22. In this case Γ is bipartite. If $t = 1$ then a subgroup Y of G of index 2 is 2-transitive on $\{a\} \cup \Gamma_2(a)$ of degree 15 with $Y_a = G_a = PSL(2,7)$. It follows that $Y \simeq A_7$, $G \simeq S_7$, and Γ is the incidence graph of the complementary design of points and hyperlanes of the projective space PG(3,2). If $t > 1$ it is easy to check for each possible t that the subgroup Y fixing the bipartition is primitive on $\Delta = \{a\} \cup \Gamma_2(a) \cup ... \cup \Gamma_{t+2}(a)$. Now Y_a is 2-transitive on $\Gamma_{t+2}(a)$ of degree 7 and by Cameron [1] Y_a should have an orbit in Δ of length greater than 14 which is not so.

In case (b), Γ is antipodal and its derived graph is $K_{8,8}$ (if $t = 1$), or a graph of type (a) (if t is odd and $t > 1$, and from the previous paragraph t must be 3 in this case), or a graph satisfying our assumptions with G_a-orbits on $V\Gamma$ of length $1,8,14,...,14$ (if t is even). We showed above that the last case cannot occur. If $t = 3$ then Γ satisfies part (e) of the theorem. So assume that $t = 1$. Then $|V\Gamma| = 32$ and Γ is bipartite. The subgroup Y of index 2 in G fixing the bipartition is imprimitive on $\{a\} \cup \Gamma_2(a) \cup \Gamma_4(a)$ with 8 blocks of length 2. If Y' is the set stabilizer of the block $\{a\} \cup \Gamma_4(a)$, then Y' contains $G_a = PSL(2,7)$ as a normal subgroup of index 2. Now G_a is 2-transitive on the remaining 7 blocks, and as the group PGL(2,7) has no representation of degree 7 it follows that $Y' = G_a \times Z$, where Z is cyclic of order 2 and Z fixes all 8 blocks setwise. In fact Z is the centre of Y and $Y/Z \simeq AGL(3,2)$. Furthermore, Z and $O_2(Y)$ are both normal in G; $O_2(Y)/Z$ is a normal elementary abelian subgroup of G/Z of order

8 and $G_a Z/Z \simeq PSL(2,7) \simeq GL(3,2)$ acts faithfully and naturally by conjugation on $0_2(Y)/Z$. It follows that G has a normal subgroup N of order 32 containing $0_2(Y)$ such that N/Z is the centralizer of $0_2(Y)/Z$ in G/Z. To complete the proof of the theorem we show that Γ has no group of automorphisms with these properties. This proof is due to Laci Kovacs.

The vertices of Γ are in 1-1 correspondence with the elements of N and the actions of $X = G_a$ on $V\Gamma$ and on N by conjugation are equivalent. Thus X has orbits on N of lengths $1,1,8,8,14$ and so the only X-invariant subgroups of N have orders $1,2,16,32$ and they are precisely $1,Z,0_2(Y)$ and N respectively. Also each orbit of length 8 generates N. Let Frat N be the frattini subgroup of N. If a subgroup of X of order 7 acted trivially on $N/$Frat N it would act trivially on N which is not the case here. It follows that $|N/$Frat $N| \geq 8$ and as Frat N is X-invariant it is either 1 or Z.

If N is not abelian then Frat $N = Z$, and also $N' = Z(N) = Z$ (for $N/Z(N)$ cannot be cyclic). Thus N is extra special and so is the central product $Q_8 * Q_8$ or $Q_8 * D_8$. However these groups have 19 and 11 elements of order 2 respectively, and so X could not act with the required orbit lengths on such a group. So N is abelian. If Frat $N = 1$ then N is elementary abelian. If N were decomposable as an X-module, $(X \simeq GL(3,2))$, it would be the direct sum of a one-dimensional and a 4-dimensional submodule, (as X acts nontrivially on N and the only nontrivial X-invariant subgroups have orders 2 and 16). Thus $N \simeq Z \times 0_2(Y)$ which is impossible as $Z \leq 0_2(Y)$. Thus N is indecomposable; a similar argument shows that N/Z is also indecomposable as an X-module, and hence N is uniserial. However as X has an orbit of length 8 in N and this orbit generates N, N is a homomorphic image of the X-module M of order 2^8 affording the permutation representation of X of degree 8. The kernel of this homomorphism would be an X-submodule of M of order 2^3, which would contain the unique minimal 1-dimensional X-submodule of M. It would follow that X would act trivially on this kernel which is not the case. Thus Frat $N \neq 1$.

Hence Frat $N = Z$ and so $N \simeq Z_2^3 \times Z_4$, and the X-invariant subgroups, written additively, are precisely, 0, $Z = 2N$, $0_2(Y) = \{y \in N; 2y = 0\} = \Omega N$, and N. An orbit of X in N of length 8 consists of elements of order 4 and generates N and so N is a homomorphic image of an X-module $W = Z_4^8$ affording the permutation representation of X of degree 8, say $N \simeq W/V$. Now W has a unique submodule U such that $W/U \simeq Z_4$ is a trivial X-module. As $W/2W \simeq Z_2^8$ clearly $2W \not\subseteq U$ and $U + 2W$ is the unique submodule S containing U with $W/S \simeq Z_2$. Further as $N \simeq W/V$ has no quotient module isomorphic to Z_4, $V \not\subseteq U$ so

$U \subset U + V$; and as $N/\Omega N$ is the unique quotient module of N on which X acts trivially, $\Omega N \simeq (U + V)/V$. Hence $U + V = U + 2W$ and $|V : U \cap V| = 2$. Also $(U \cap V) + 2W$ is either equal to, or has index 2 in $V + 2W$. Now $(V + 2W)/V \simeq Z$ has order 2, so in the former case $2W/(2W \cap (U \cap V)) \simeq (V + 2W)/(U \cap V)$ is a 2-dimensional quotient module of $2W$ on which X must act trivially, and $2W$ (which is Z_2^8 affording the permutation representation) has no such quotient. In the latter case $((U \cap V) + 2W)/2W$ is a submodule of $W/2W$ of order $|(V + 2W)/2W|/2 = 8$, and $W/2W$ (which is also Z_2^8 affording the permutation representation of X of degree 8) has no such submodule. Thus Γ has no group of automorphisms with the required properties and the proof of the theorem is complete.

Proof of Corollary. Let G be primitive on a set Ω and suppose that for $a \in \Omega$, G_a has an orbit $\Gamma(a)$ such that $PSL(2,q) \leq G_a^{\Gamma(a)} \leq P\Gamma L(2,q)$ of degree $q + 1$. By [1], $\Gamma \circ \Gamma^*(a)$ is an orbit of G_a of length $q(q + 1)/k$ where $k \leq q/2$. If G_a is unfaithful on $\Gamma(a)$ then $k = 1$ by [7]. So assume that G_a is faithful on $\Gamma(a)$. By an argument similar to that in the first paragraph of the proof of the theorem, k is 1 or 2.

REFERENCES

1. P. J. Cameron, Permutation groups with multiply transitive suborbits, Proc. London Math. Soc. (3) 25 (1972), 427–440.

2. P. J. Cameron, Suborbits in transitive permutation groups, in: Combinatorial Group Theory, Math. Centre Tracts 57, Amsterdam 1974, Edited by M. Hall Jr. and J. H. Van Lint, 98–129.

3. P. J. Cameron, Finite permutation groups and finite simple groups, (to appear).

4. P. J. Cameron and C. E. Praeger, Graphs and permutation groups with projective subconstituents, J. London Math. Soc. (to appear).

5. L. E. Dickson, Linear Groups with an Exposition of the Galois Field Theory, Dover, New York, 1958.

6. A. Gardiner, Symmetry conditions in graphs, in: Surveys in Combinatorics, London Math. Soc. Lecture Notes 38, Cambridge University Press, Cambridge, 1979, Edited by B. Bollobas, 22–43.

7. C. E. Praeger, Primitive permutation groups and a characterization of the odd graphs, J. Combinatorial Theory Series B.

8.   C. E. Praeger, Symmetric graphs and a characterization of the odd graphs, in: Combinatorial Mathematics VII, Ed. R. W. Robinson, G. W. Southern and W. D. Wallis, Lecture Notes in Math. No. 829, Springer, Berlin, Heidelberg, New York, 1980, 211–219.

9.   C. C. Sims, Computational methods in the study of permutation groups, in: Computational Problems in Abstract Algebra, Pergamon Press, London, 1970, Edited by J. Leech, 169–183.

10. M. Suzuki, Transitive extensions of a class of doubly transitive groups, Nagoya Math. J. 27 (1966), 159–169.

11. H. Wielandt, Finite Permutation Groups, Academic Press, New York, 1964.

THE GROUP OF THE HUNGARIAN MAGIC CUBE

CHRIS ROWLEY

Faculty of Mathematics
The Open University
Milton Keynes
England

1. INTRODUCTION

The Büvös Kocka, which is more commonly called the Hungarian Magic Cube after
its country of origin, is a cube which appears to be split, in the natural
way, into 27 identical small cubes which are mechanically linked in such a
way that any layer of 9 small cubes in any plane parallel to a face of the
large cube can be rotated in that plane through any angle, but these are
essentially the only possible movements (up to rigid movements of the whole
cube). In this paper we shall investigate the group of all the configurations which can be achieved by using only these rotations and in which the
small cubes still fit together in a natural way to form a large cube. Since
we are interested only in the configurations relative to the centre of the
large cube, we take the central small cube to be fixed both in position and
orientation. The configurations can then all be achieved using the 6 rotations through a quarter revolution (in a fixed direction) of each of the
six layers of 9 small cubes which contain a face of the large cube; we call
these six rotations *the generators*.

The practical problem of how to actually achieve any prescribed, and attainable, configuration using the generators has been attacked vigorously by many with more patience than the present author, and many results are contained in [4] and [5]. Their efforts do raise one group-theoretic problem: given a group H and a set S of generators of H, what is the maximum over all $h \in H$ of the minimum length of h as a word in the generators in S, and how can we find an economical expression for a given element in H as a word in these generators?

All the results used in this paper on attainability of certain configurations, which are in the appendix, are in [5]; they have been discovered by a large and rapidly growing number of "cubists", to whom my thanks are due, together with Professor Martin Isaacs for some useful discussions on the detailed group-theoretic arguments, Don Taylor for arousing my interest in the group theory of the cube through his notes on the subject [6] and David Singmaster for his unflagging concern for the beast.

The group (which we shall label G) of configurations which we wish to investigate is not, a priori, very clearly defined: we have a set of 6 generators for it and we can discover relations between them, but there is no way of knowing when we have found a defining set of relations. We therefore need to find first a set on which the group has a faithful action; we can then specify G precisely as a subgroup of the symmetric group on this set. In Section 2 we find a suitable set of size 72; we can then pose the problem more rigorously as that of finding the subgroup of S_{72} generated by a certain set of 6 permutations. However, this would not be a very interesting problem, nor would the answer be very illuminating, so it is best to approach the problem as an application of group theory to a problem in the real-world, and this is the spirit in which this paper is written.

Notation and Conventions

The following notation and conventions are additional to those described in the introduction and to the standard group theory notation, which is that of Gorenstein's book [1] or, for permutation groups, Wielandt's book [7].

The *Cube* (capital c) refers to the whole object, which we consider to be simply the set of 27 small cubes, each of which is in turn merely the set of its 6 faces.

The *original position* of the Cube is a distinguished configuration to which all the other configurations are compared in order to identify them. An actual Cube is decorated in such a way that this identification is straightforward.

A *generator* is a change of configuration which is given by a twist through a quarter revolution (in a fixed direction) of a layer of cubes which form a face when we think of the Cube as an actual object in 3-space. Each generator has order 4.

Each element of the group G is, by definition, a product of the generators. It seems to be standard amongst cubists to use the following non-standard definition of the length of an element, so we shall follow their convention in this paper. We define the *length of the word* $\prod_{i=1}^{n} g_i^{r_i}$ to be n, where for $i = 1,\ldots,n$, g_i is a generator, $g_i \neq g_{i+1}$, and $r_i \neq 0 \pmod 4$. Then we define the *length of an element* $g \in G$ to be the minimum of the lengths of all those words (in the generators) which are equal to g.

The set of 27 small cubes is the disjoint union of the following subsets (See Figure 1):

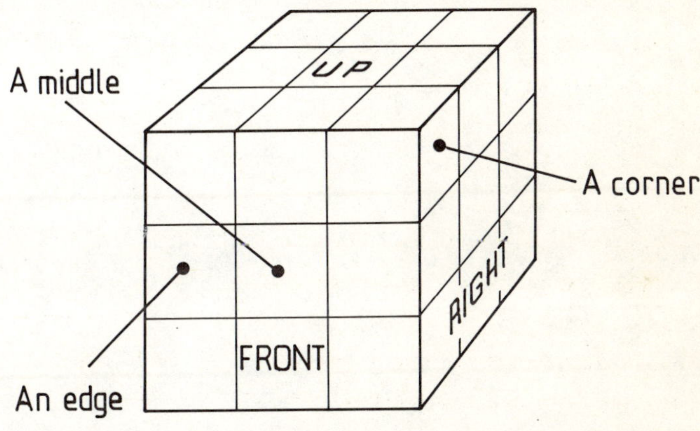

Figure 1

(a) E', the set of small cubes which are in the middle of an edge of the Cube and so have only two faces which are external to the Cube - there are 12 elements in E' and we shall call these *edges*;

(b) C', the set of small cubes which form corners of the Cube and so have three faces which are external to the Cube - there are 8 elements in C' and we shall call these *corners*;

(c) 6 singletons, each consisting of a small cube in the middle of one of the 6 faces of the Cube and so having only one face external to the Cube - these 6 elements are called *middles* and labelled M_i', $i = 1,\ldots,6$ (we shall also often use the label M_i' for the set $\{M_i'\}$);

(d)   another singleton which contains the unique small cube which has no faces external to the Cube - this small cube is called the *central cube*.

Let Ω be a finite set; we shall write $\text{Sym}(\Omega)$ and $\text{Alt}(\Omega)$ respectively for the symmetric group and alternating group on Ω.   If H is a subgroup of $\text{Sym}(\Omega)$ and Δ is a subset of Ω, then

$$H_\Delta = \{h \in H : \delta^h = \delta \quad \text{for all} \quad \delta \in \Delta\}$$

$$H_{\{\Delta\}} = \{h \in H : \delta^h \in \Delta \quad \text{for all} \quad \delta \in \Delta\}$$

$H_{\{\Delta\}}^{\Delta} = H_{\{\Delta\}}/H_\Delta$ and we shall consider this to be a subgroup of $\text{Sym}(\Delta)$. Suppose that H is a subgroup of $\text{Sym}(\Omega_1)$ and K is a subgroup of $\text{Sym}(\Omega_2)$.   If $\Omega_1 \cap \Omega_2 = \phi$ then we shall consider $H \times K$ to be the subgroup of $\text{Sym}(\Omega_1 \cup \Omega_2)$ given by

$$w^{(h,k)} = \begin{cases} w^h & \text{if } w \in \Omega_1 \\ w^k & \text{if } w \in \Omega_2 \end{cases}$$

for all $w \in \Omega$ and $(h,k) \in H \times K$.   By $H \wr K$ we shall mean the permutation wreath product of H by K, as in the introduction of [2], which is a subgroup of $\text{Sym}(\Omega_1 \times \Omega_2)$.

Finally note that if $H \cong Z_4$ and $|\Omega| = 4$ then there is, up to isomorphism, only one faithful action of H on Ω, so we can denote this situation by $H = Z_4 \subseteq \text{Sym}(\Omega)$.

2.   THE SET

Since the Cube is made up of 27 small cubes and we are taking the central cube to be fixed, an element of the group G will be completely determined by a precise description of the configuration of the 26 outer cubes relative to the central cube.   Each generator maps each of these 26 cubes into the position previously occupied by another of them and, thus, defines a bijection of this set of 26 cubes.   Each generator also maps edges to edges, corners to corners and fixes, as a whole, each of the 6 middles.   Therefore the set E' of 12 edges, the set C' of 8 corners and the 6 middles $M_1', \ldots, M_6'$ must be G-invariant subsets of this set of 26 cubes.   (Here we are identifying each M_i' with the singleton $\{M_i'\}$.)

It is not however sufficient to consider merely the action of G in permuting these 26 cubes because it is possible, using the generators, to restore a small cube to its original position but in a different orientation. Each generator, and hence the whole of G, does however map the small cubes into a configuration in which those faces of the small cubes which were part of the faces of the large Cube in the original position retain this property. Therefore G acts as a group of bijections of these $9 \times 6 = 54$ small faces which make up the 6 faces of the large Cube.

Since E' and C' are G-invariant, so are E and C, where E is the set of 24 small faces of small cubes in E' and C is the set of 24 small faces of small cubes in C'. We shall call elements of E *edge faces* and elements of C *corner faces*. The positions occupied by each of the elements of $E \cup C$ completely determine the position and orientation of each cube in $E' \cup C'$. The position of each middle is fixed and so is the position of the one face of that middle which is part of a face of the large Cube, but this face can be turned through each of the 4 distinct multiples of a quarter revolution twist. This is achieved by the 4 distinct powers of the unique generator which turns that middle. Thus, for $i = 1,\ldots,6$, G acts on the set of four orientations of M_i' (or, equivalently, the four vertices of its outer face); we call this set M_i.

3. THE GROUP

We have now established that each of the configurations of the Cube relative to the central cube can be uniquely described by a permutation of $E \cup C \cup M$, where $M = \bigcup_{i=1}^{6} M_i$. Therefore in the next section we shall describe G as a subgroup of $\mathrm{Sym}(E \cup C \cup M)$ or, more precisely, since E, C and M are G-invariant, as a subgroup of $\mathrm{Sym}(E) \times \mathrm{Sym}(C) \times \mathrm{Sym}(M)$.

Since

$$|E| = 12 \times 2 = 24$$

$$|C| = 8 \times 3 = 24$$

and

$$|M| = 6 \times 4 = 24$$

this means that, as an abstract group, G is isomorphic to a subgroup of S_{72} or, more precisely, to a subgroup of $S_{24} \times S_{24} \times S_{24}$. We shall need to continue to consider the actions of G on E' and C', where now each element $B \in E' \cup C'$ is identified with the set of elements of $E \cup C$ which are the

outer faces of B. Thus E is now the disjoint union of the elements of E'
and C is the disjoint union of the elements of C'. We shall also identify
E with $X \times E'$, where $X \in E'$ and C with $Y \times C'$, where $Y \in C'$.

Since E' is a block system for G^E, this action is contained in the full
automorphism group of this block system, which is the wreath product $E^* =$
$Sym(X) \wr Sym(E')$. A similar argument plus a little geometry shows that
$G^{(C)} \subseteq C^* = Alt(Y) \wr Sym(C')$. Thus $G \subseteq E^* \times C^* \times G^M$. We shall prove that
G has very small index in this group – namely 24, and also that $G^M \cong Z_4^6$. We
shall first determine precisely the actions of G on E, C and M; then we shall
investigate the embedding of G in $G^E \times G^C \times G^M$ to determine G completely.

We shall prove the following.

THEOREM. The group G is the unique subgroup of index 2 in

$$((E^* \cap Alt(E)) \times 0^3(C^*) \times \prod_{i=1}^{6} Z_4^{(i)}) \cap Alt(E \cup C \cup M)$$

such that $G^{E' \cup C'} = (Sym(E') \times Sym(C')) \cap Alt(E' \cup C')$, where $Z_4^{(i)} = G^{M_i} \cong$
$Z_4 \subseteq Sym(M_i)$.

NOTE 1. Since $|Sym(X)| = 2$ and $|Sym(E')| = 12!$, we have $|E^*| = 2^{12} \times 12!$
and so $|E^* \cap Alt(E)| = (2^{12} \times 12!)/2$. Also $|Alt(Y)| = 3$ and $|Sym(C')| = 8!$
so that $|0^3(C^*)| = (3^8 \times 8!)/3$. Finally $|\prod_{i=1}^{6} Z_4^{(i)}| = 4^6$ so

$$|G| = \frac{1}{2} \times \frac{1}{2} \times \frac{2^{12} \times 12!}{2} \times \frac{3^8 \times 8!}{3} \times 4^6$$

Here one of the factors 1/2 comes from the restriction to $Alt(E \cup C \cup M)$
and the other from the restriction on $G^{E' \cup C'}$ to $Alt(E' \cup C')$. Numerologists
may be interested to know that

$$|G| = 2^{38} \times 3^{14} \times 5^3 \times 7^2 \times 11 = 885\ 80102\ 70615\ 52250\ 88000$$

NOTE 2. The group that is often referred to as the "group of the cube" is
in fact the action $G^{E \cup C}$. This is because actual Cubes are not normally
decorated in such a way that the action on M is detectable.

The proof of the theorem consists of the following sequence of lemmas.

LEMMA 3.1. Let $X \in E'$. Then

(1) $G^X_{\{X\}} = Sym(X)$

(2) $G^{E'} = Sym(E')$

and

(3) $G^E \subseteq E* = Sym(X) \wr Sym(E')$

Proof.   The relationship defined by:

a ~ b if and only if there is a $B \in E'$ such that $a,b \in B$

is a G-congruence on E with quotient E'.

(1) For any two elements of E there is an element of G of length at most 3 which maps one to the other, so that G acts transitively on E and hence on E'. Since $|X| = 2$, this immediately gives $G^X_{\{X\}} = Sym(X)$.

(2) The stabilizer of an edge $B \in E'$ is transitive on $E' - \{B\}$ (elements of G of length at most 3 again demonstrate this). There is also an element g_1 in G whose action on E' is a single 3-cycle. Thus, by a result of C. Jordan in [3], $G^{E'} \geq Alt(E')$. Now each generator gives an odd permutation on E' so $G^{E'} = Sym(E')$.

(3) This follows from arguments like those in the introduction to [2].

LEMMA 3.2. $G^E = E* \cap Alt(E) = (Sym(X) \wr Sym(E')) \cap Alt(E)$.

Proof.   Each generator gives an even permutation of E so that $G^E \subseteq E* \cap Alt(E)$, which has index 2 in E*. There is an element $g_2 \in G$ which is the identity on E' and, in E, acts nontrivially on the elements of exactly two of the elements of E'. Therefore, since the action on E' is 2-transitive, G contains sufficient conjugates of this element to ensure that G^E has index at most 2 in E*. Thus $G = E* \cap Alt(E)$.

LEMMA 3.3. Let $Y \in C'$. Then

(1) $G^Y_{\{Y\}} = Alt(Y)$

(2) $G^{C'} = Sym(C')$

and

(3) $G^C \subseteq C* = Alt(Y) \wr Sym(C')$

Proof.   The argument here is much the same as in Lemma 3.1: we have a G-congruence on C with quotient C'; we can map a given element of C to any

of the 24 attainable positions (using an element of length at most 2); if
$B \in C'$ then G_B is transitive on $C' - \{B\}$ (again using elements of length at
most 2); and there is an element $g_3 \in G$ which gives a single 3-cycle on C'.
Finally, each generator also gives an odd permutation on C', proving (2).

To complete the proof of (1) we must show that only the three even
permutations of Y can arise. This is because the generators all move the
small cubes rigidly in 3-space and so the permutations of the six faces of
a small cube produced by those elements of G which fix it as a whole, must
be in the group H of 24 permutations produced by the isometries of the
cube whose determinant is +1. The setwise stabiliser in H of the set Y of
three pairwise adjacent faces acts on the elements of Y as Alt(Y) and hence,
since $G_{\{Y\}}$ acts transitively on Y, we have $G_{\{Y\}}^{Y} = $ Alt(Y).

Again (3) follows immediately from (1) and (2).

LEMMA 3.4. $G^C = 0^3(C*) = 0^3(\text{Alt}(Y) \wr \text{Sym}(C'))$.

Proof.   Let L be the setwise stabiliser of Y in C*. Then $L = M \times L_Y$,
where $M \cong \text{Alt}(Y)$. Let $\phi : C* \rightarrow L/L_Y$ be the transfer homomorphism relative
to L and the canonical homomorphism from L onto L/L_Y, as defined in Section
7.3 of [1]. Let $K = \ker\phi$, then $M \not\subseteq K$ and so ϕ is surjective. However, G
is generated by elements of order 4 so $G^C \subseteq K$.

The next part of the proof is similar to that of Lemma 3.2: there is
an element $g_4 \in G$ which is the identity on C' and acts nontrivially on the
elements of exactly two of the elements of C', turning them in opposite
directions (i.e. the action on one is not conjugate in the group C* to the
action on the other); transitivity on C' therefore shows that G^C has index
at most 3 in C* giving $G^C = K$.

That $K = 0^3(C*)$ is proved as follows. Let $\tilde{H} = (C*)^{(C')}$ and $\bar{H} =$
$C*/0^3(C*)$. The element g_4 above is in $[C*,C*']$ and $[\tilde{H},\tilde{H}'] = \tilde{H}'$, therefore
$[C*,C*'] = C*'$. Thus $[\bar{H},\bar{H}'] = \bar{H}'$ and so $\bar{H}' = 1$. Now $|\tilde{H}/\tilde{H}'| = 2$ giving
$|\bar{H}| = 3$.

LEMMA 3.5. $G^M = \prod_{i=1}^{6} Z_4^{(i)}$, where $Z_4^{(i)} = G^{M_i} = Z_4 \subseteq \text{Sym}(M_i)$.

Proof.   The argument of the penultimate paragraph of Section 2 shows
that, for each $i \in \{1,\ldots,6\}$, there is a unique generator which acts as $Z_4^{(i)}$
on M_i and fixes each element of M_j for $j \neq i$. This lemma is then merely an
example of our conventions concerning the notation for the action of a group
on a union of sets.

We next look at the action of G on $E' \cup C'$ in order to determine $G^{E \cup C}$ as a subgroup of $G^E \times G^C$.

LEMMA 3.6. $G^{E' \cup C'} = (\mathrm{Sym}(E') \times \mathrm{Sym}(C')) \cap \mathrm{Alt}(E' \cup C')$.

Proof. Since the elements g_1 and g_3 used in the proofs of Lemmas 3.1 and 3.3 are in G_C and G_E respectively, we have $G^{E' \cup C'} \supseteq \mathrm{Alt}(E') \times \mathrm{Alt}(C')$. Now each generator gives an odd permutation of both E' and C' but an even permutation of $E' \cup C'$. Since $(\mathrm{Sym}(E') \times \mathrm{Sym}(C'))/(\mathrm{Alt}(E') \times \mathrm{Alt}(C')) \cong Z_2 \times Z_2$, this proves that $G^{E' \cup C'}$ is the above subgroup.

LEMMA 3.7. Let A^E be the subgroup of G^E which is the inverse image of $\mathrm{Alt}(E')$ under the homomorphism defined by the action of G^E on E' and let $A^C \subseteq G^C$ be similarly defined as the inverse image of $\mathrm{Alt}(C')$. Then $G^{E \cup C}$ is the unique group H such that

$$|G^E \times G^C : H| = 2$$
$$A^E \times A^C \subseteq H \subseteq G^E \times G^C$$

and

$$H^{E' \cup C'} = G^{E' \cup C'}.$$

Proof. The elements g_2 and g_4 used in the proofs of Lemmas 3.2 and 3.4 are in G_C and G_E respectively so the arguments of those lemmas together with Lemma 3.6 show that $A^E \times A^C \subseteq G^{E \cup C}$. Since the kernel of the action of $G^{E \cup C}$ on $E' \cup C'$ is contained in $A^E \times A^C$, the action on $E' \cup C'$ now defines $G^{E \cup C}$ completely as a subgroup of index 2 in $G^E \times G^C$.

LEMMA 3.8. $G = (G^{E \cup C} \times G^M) \cap \mathrm{Alt}(E \cup C \cup M)$.

Proof. Since each generator gives an even permutation on $E \cup C \cup M$, we have $G \subseteq \mathrm{Alt}(E \cup C \cup M)$. Now $G/(G_{E \cup C}G_M)$ is isomorphic to a factor group of $G/G_{E \cup C} = G^{E \cup C}$ and to a factor group of $G/G_M = G^M$. Also G^M is abelian and $|G^{E \cup C}/(G^{E \cup C})'| = 2$, therefore $|G/(G_{E \cup C}G_M)| \leq 2$. If this index were 1 then, since $G_{E \cup C} \cap G_M = 1$, we should have $G = G^{E \cup C} \times G^M \subseteq \mathrm{Alt}(E \cup C \cup M)$. This contradiction shows that

$$|G^{E \cup C} \times G^M : G| = |G/(G_{E \cup C} \times G_M)| = 2$$

Thus

$$G = (G^{E \cup C} \times G^M) \cap \mathrm{Alt}(E \cup C \cup M)$$

This concludes the proof of our theorem but leaves a few further points of interest concerning the structure of G, whose proofs we outline.

PROPOSITION. (1) G splits over $G_M = A^E \times A^C$ (in the notation of Lemma 3.7).
  (2) G does not split over G_E, G_C or G_{EUC}.

Proof.   (1) Let x be a fixed element of G^E which is a product of 2 transpositions. The multiple transitivity of G^E implies the existence of elements h_i, $i = 1,\ldots,6$, such that

$$h_i = (x,1,k_i) \in G^E \times G^C \times G^M$$

where k_i, $i = 1,\ldots,6$, are the images in G^M of the generators. Then $\langle h_i | i = 1,\ldots,6 \rangle$ is a complement of G_M.

  (2) These follow from the fact that if $x \in G$ has order 2, then $xG_M \in \Omega_2(G^M)$ and this is contained in $(G_{EUC}G_M)/G_M$. Therefore $x \in G_{EUC}$.

4. APPENDIX

Here we show how to obtain certain configurations referred to in Section 3 using the generators. To describe the group elements we label the six generators, f, b, r, ℓ, u and d; the faces are labelled front, back, left, right, up and down in a natural way, that is, such that (front, back), (left, right) and (up, down) are the three pairs of nonadjacent faces and front, up, right meet at a vertex in this cyclic order going clockwise looking at the vertex from the outside (as in Figure 1); and each generator rotates the cubes forming the face whose label begins with that letter. These labellings are such that r moves the corner at the vertex where front, up, right meet to the vertex where up, back, right meet and such that, if xylophone and yellow are any two adjacent faces, then xy fixes the position of one of the corners occupying the two vertices which are common to xylophone and yellow. Also, vz means perform first v and then z.

  (1) $g_1 = b^{-1}u^{-1}b\ell frur^{-1}f^{-1}\ell^{-1}$ fixes every element of C ∪ M and all but six elements of E. On E' it is the 3-cycle (up-front, up-right, back-up).
  (2) $g_2 = u^{-1}fr^{-1}uf^{-1}r^{-1}\ell df^{-1}rd^{-1}fr\ell^{-1}$ fixes every element of C ∪ M and all but four elements of E. On E' it is the identity but it changes the orientation of front-up and front-down.

(3) $g_3 = b\ell^{-1}b^{-1}rb\ell b^{-1}r^{-1}$ fixes every element of E $\cup$ M and all but nine elements of C. On C' it is the 3-cycle (front-left-up, right-up-back, right-front-up).

(4) $g_4 = (\ell b^2\ell^{-1}uf^2u^{-1})^2$ fixes every element of E $\cup$ M and all but six elements of C. On C' it is the identity but it changes the orientation of right-up-back and down-left-front.

REFERENCES

1. D. Gorenstein, Finite Groups, Harper and Row, New York, 1968.

2. W. C. Holland, The Characterisation of Generalised Wreath Products, J. Algebra 13 (1969), 152-172.

3. C. Jordan, Théorèmes sur les groupes primitifs, Math. Pures et Appl. 16 (1871), 383-408.

4. K. Ollerenshaw, The Hungarian Magic Cube, Bull. Inst. Maths. Applics. 16 (1980), 86-92.

5. D. Singmaster, Notes on the "Magic Cube", David Singmaster, London, 1979.

6. D. E. Taylor, Mastering Rubik's Cube, Book Marketing Australia, Richmond, 1980.

7. H. Wielandt, (transl. R. Bercov), Finite Permutation Groups, Academic Press, New York, 1964.

DUAL ABELIAN GROUPS IN THE DESIGN OF
EXPERIMENTS

R. A. BAILEY*

Department of Mathematics
The Open University
Milton Keynes
England

1. A TOMATO EXPERIMENT

Consider an experiment to determine which of t varieties of tomato plant,
grown in which of c types of compost, produces the greatest yield in weight
of fruit per plant. The items under test, called *treatments*, are all tc
combinations of variety of tomato with type of compost. The varieties of
tomato are called *levels* of *treatment factor* "Tomatoes", while the types
of compost are called levels of treatment factor "Compost". We shall abbre-
viate these treatment factors to T and C respectively.

An experimental unit which receives a particular treatment is called a
plot. In this case, a plot is simply a position within a greenhouse where
a particular tomato plant is grown in a particular compost. We shall suppose
that each treatment is tested just once in the experiment, so that there are
tc plots.

Current affiliation: Statistics Department, Rothamsted Experimental
Station, Harpenden, England.

When the experiment has been performed and the yields recorded, treatments are compared. Such comparisons are made by calculating linear combinations of the yields: for example, the difference between the average yields of two varieties of tomato plants. The set of all such linear combinations forms a Euclidean vector space $\mathbb{R}^{tc}$ with orthonormal basis corresponding to the plots. We distinguish four subspaces of $\mathbb{R}^{tc}$: namely $V_0 = \{v \in \mathbb{R}^{tc}:$ all coordinates of v are the same$\}$; $V_T = \{v \in \mathbb{R}^{tc}:$ if two plots receive the same level of T then corresponding coordinates of v are equal$\}$, with a similar definition for V_C; and finally $V_{TC} = \{v \in \mathbb{R}^{tc}:$ if two plots receive the same level of T and of C then corresponding coordinates of v are equal$\} = \mathbb{R}^{tc}$.

In general, the only linear combinations used for estimation are those orthogonal to V_0: these are called *contrasts*. In this paper we shall use $V_1 = V_2 \oplus V_3$ to mean that V_1 is the *orthogonal* direct sum of V_2 and V_3, and $V_1 = V_2 - V_3$ to mean that V_1 is the orthogonal complement of V_3 in V_2.

Evidently $V_0 \leq V_T \leq V_{TC}$ and $V_0 \leq V_C \leq V_{TC}$. We define four further subspaces of $\mathbb{R}^{tc}$ by:

$$W_0 = V_0$$
$$W_T = V_T - V_0$$
$$W_C = V_C - V_0$$
$$W_{TC} = V_{TC} - \langle V_T, V_C \rangle$$

where $\langle \ldots \rangle$ denotes "subspace generated by $\ldots$".

LEMMA 1. The spaces W_T and W_C are orthogonal.

Proof.   For each level i of T, denote by a_i the vector whose ω-th coordinate is 1 if plot ω receives level i of T, and 0 otherwise. Similarly, define a vector b_j for each level j of C. Then $a_i \cdot b_j = 1$. Vectors in W_T have the form $\sum_i \alpha_i a_i$, where $\alpha_i \in \mathbb{R}$ and $\sum \alpha_i = 0$. Similarly, vectors in W_C have the form $\sum_j \beta_j b_j$, where $\beta_j \in \mathbb{R}$ and $\sum \beta_j = 0$. Now,

$$\left(\sum_i \alpha_i a_i\right) \cdot \left(\sum_j \beta_j b_j\right) = \sum \sum \alpha_i \beta_j = \left(\sum \alpha_i\right)\left(\sum \beta_j\right) = 0$$

Since every other pair of W-spaces is orthogonal by construction, Lemma 1 shows that $\mathbb{R}^{tc} = W_0 \oplus W_T \oplus W_C \oplus W_{TC}$. Moreover, the dimensions of the subspaces so far discussed are:

subspace	V_0	V_T	V_C	V_{TC}	W_0	W_T	W_C	W_{TC}
dimension	1	t	c	tc	1	t − 1	c − 1	(t − 1)(c − 1)

In the statistical literature these dimensions are called the *numbers of degrees of freedom*. The spaces W_0, W_T, W_C, W_{TC} are called, respectively, the *mean*, the *main effect* of T, the main effect of C, and the TC-*interaction*. For a good account of the practical rationale behind this decomposition of $\mathbb{R}^{tc}$, see [8], Chapter VI.

2. GENERAL FACTORIAL TREATMENT STRUCTURE

In a general factorial experiment there are n treatment factors, $A_1, \ldots, A_n$. Denote by Ω_i the set of m_i levels of A_i: it is assumed that $m_i \geq 2$ for each i. Then the set of treatments is $\Omega = \Omega_1 \times \ldots \times \Omega_n$. If each treatment is tested on just one plot we may identify Ω with the set of plots. It is then convenient to regard the vector space of linear combinations as $\mathbb{R}^{\Omega}$, the vector space of all functions from Ω to $\mathbb{R}$, with inner product defined by

$$f \cdot g = \sum_{\omega \in \Omega} (\omega f) \times (\omega g)$$

for $f, g \in \mathbb{R}^{\Omega}$: here, as throughout this paper, functions are written on the right.

For $I \subseteq \{1, \ldots, n\}$, denote by Ω_I the set $\prod_{i \in I} \Omega_i$, this being taken to mean a singleton when $I = \emptyset$. Moreover, denote by π_I the natural projection from Ω onto Ω_I. Then the analogue of the V- and W- subspaces defined in §1 are

$$V_I = \{f \in \mathbb{R}^{\Omega}: \text{ there exists } f' \in \mathbb{R}^{\Omega_I} \text{ such that } f = \pi_I f'\}$$

and

$$W_I = V_I - \langle V_J: J \subseteq I, J \neq I \rangle$$

where

$$\dim(V_I) = \prod_{i \in I} m_i$$

and

$$\dim(W_I) = \prod_{i \in I} (m_i - 1)$$

and moreover

$$\mathbb{R}^{\Omega} = \bigoplus_{I \subseteq \{1, \ldots, n\}} W_I$$

is an orthogonal decomposition of $\mathbb{R}^{\Omega}$ into subspaces which have a definite meaning in terms of the treatment structure.

3. BLOCKING

In the tomato experiment described in §1, a single greenhouse may not be
sufficient to contain all the varieties of tomato plant in every type of
compost. In a more general experiment, if the set of plots available for
use does not contain a subset of sufficiently many similar plots then we
may have to partition the set of plots used into *blocks*, so that the plots
within each block are more or less homogeneous. However the difference
between blocks is a nuisance factor in the experiment. For example, if each
greenhouse contains tomato plants of only one variety, there will be no way
of distinguishing between variety differences and greenhouse differences.

In general, we define a subspace V_{BLOCKS} of $\mathbb{R}^{\Omega}$ in an analogous way to
the V_I. Clearly, $V_{BLOCKS} \supseteq V_{\emptyset}$. If $V_{BLOCKS} \supseteq V_{\{i\}}$ for any $i \in \{1,\ldots,n\}$
then differences due to treatment factor A_i are indistinguishable from block
differences: we say that the main effect of A_i is *confounded* with blocks.
Similarly, if $I \subseteq \{1,\ldots,n\}$ and $|I| \geq 2$, the I-interaction is confounded
with blocks if $W_I \subseteq V_{BLOCKS}$. If each treatment is tested on a single plot,
there must obviously be some confounding if there are blocks in the experiment. Usually the contrasts in W_I are considered less important the larger
W_I is, so it is desirable to confound W_I for large subsets I only. However,
this is not usually possible, as the following Lemma shows.

LEMMA 2. If $i \in I \subseteq \{1,\ldots,n\}$ and $W_I \subseteq V_{BLOCKS}$ and $m_i \geq 3$ then $V_{\{i\}} \subseteq$
V_{BLOCKS}.

Proof.   For each $j \in I\backslash\{i\}$ we can find an element $f_j \in W_{\{j\}}$ such that,
for all $\omega \in \Omega$, $\omega f_j \neq 0$. Let $\alpha \in \Omega_i$. Since $m_i \geq 3$ we can find an element
$f_i \in W_{\{i\}}$ such that $\omega f_i = 0$ if and only if $\omega \pi_{\{i\}} = \alpha$. Let f be the pointwise
product of f_j, $j \in I$. Then $f \in W_I$, and $\omega f = 0$ if and only if $\omega \pi_{\{i\}} = \alpha$.
Since $f \in V_{BLOCKS}$, this implies that $\alpha \pi_{\{i\}}^{-1}$ is a union of blocks. This is
true for all choices of $\alpha \in \Omega_i$, and so $V_{\{i\}} \subseteq V_{BLOCKS}$.

Thus we have a twofold problem.

(1) Find a finer decomposition of $\mathbb{R}^{\Omega} = \oplus W_I$

(2) Find a method of dividing the treatments into blocks in such a way
that V_{BLOCKS} is a direct sum of some of the new spaces into which $\mathbb{R}^{\Omega}$ is
decomposed.

We tackle (1) in the next two sections, and (2) in §6.

4. FINITE ABELIAN GROUPS AND THEIR DUALS

The sets Ω_i, being sets of levels of treatment factors, may or may not have an internal structure. However, for the purposes of solving problem (1) we identify Ω_i with a cyclic group of order m_i generated by ω_i. (We could in fact identify Ω_i with any *abelian* group of order m_i, and Theorem 1 would still be valid: however, we restrict the exposition to the case where Ω_i is cyclic.) Then Ω is the direct product of the groups Ω_i.

The spaces $V_{\{i\}}$ are determined by the i-th coordinates of elements in Ω. If Ω were a vector space the operation of picking out the i-th coordinate would be an element of the dual space. This suggests that we examine the dual group Ω^* of Ω. A clear exposition of dual groups may be found in [9], Section V.6. The dual group Ω^* consists of all irreducible characters of Ω; that is, of all homomorphisms from Ω into the multiplicative group of $\mathbb{C}$. Multiplication in Ω^* is defined pointwise. In fact Ω^* is the direct product of the cyclic groups $\langle \chi_j \rangle$, $j = 1,\ldots,n$, where χ_j is defined by

$$\omega_j \chi_j = \exp(2\pi i/m_j)$$

$$\omega_k \chi_j = 1 \text{ if } k \neq j$$

Thus if $\omega = (\omega_1^{z_1},\ldots,\omega_n^{z_n}) \in \Omega$ and $\phi = (\chi_1^{x_1},\ldots,\chi_n^{x_n}) \in \Omega^*$, we have

$$\omega\phi = \exp(2\pi i[\omega,\phi]/e) \tag{4.1}$$

where e is the exponent of Ω and

$$[\omega,\phi] = \sum_{j=1}^{n} z_j x_j e/m_j \text{ modulo } e \tag{4.2}$$

In practice it is easier to work with Equation (4.2) directly rather than with Equation (4.1). If Ω and Ω^* are written additively they may be thought of as $\mathbb{Z}_e$-modules: in this case the function $[\ ,\ ]$ defined by Equation (4.2) is a bilinear form.

By analogy with the subspaces V_I, we define, for each $x \in \Omega^*$, the subspace S_x of $\mathbb{R}^\Omega$ by

$$S_x = \{f \in \mathbb{R}^\Omega : \text{ there exists } f' \in \mathbb{R}^\vee \text{ such that } f = xf'\}$$

In other words, $f \in S_x$ if and only if f is constant on each coset in Ω of the annihilator $\text{Ann}(x)$ of x, where $\text{Ann}(x) = \{\omega \in \Omega: \omega x = 1\}$.

LEMMA 3. If x has order r then S_x has dimension r. Moreover, if $y \in \langle x \rangle$ then $S_y \subseteq S_x$.

It follows from Lemma 3 that, if $y \in \langle x \rangle$, then $S_y = S_x$ if and only if $\langle y \rangle = \langle x \rangle$. Since we require orthogonal subspaces of $\mathbb{R}^\Omega$, we further define, for each $x \in \Omega^*$, $T_x = S_x - \langle S_y : y \in \langle x \rangle, \langle y \rangle \neq \langle x \rangle \rangle$

EXAMPLE 1. If x has order 9 and $y = x^3$, we have the following subspaces and dimensions:

S-space	dimension	T-space	dimension
S_x	9	$T_x = S_x - S_y$	6
S_y	3	$T_y = S_y - S_1$	2
S_1	1	$T_1 = S_1$	1

EXAMPLE 2. If x has order 15, $y = x^3$ and $z = x^5$, we have the following subspaces and dimensions:

S-space	dimension	T-space	dimension
S_x	15	$T_x = S_x - \langle S_y, S_z \rangle$	8
S_y	5	$T_y = S_y - S_1$	4
S_z	3	$T_z = S_z - S_1$	2
S_1	1	$T_1 = S_1$	1

5. COMPLEX EXTENSION

We have so far been considering the vector space $\mathbb{R}^\Omega$ because in most practical experiments the yields are measured in real numbers, and only real linear combinations of the yields are meaningful. However, if we extend the field of scalars to $\mathbb{C}$ we achieve some simplification. The inner product is extended in the usual way; viz:

$$f \cdot g = \sum_{\omega \in \Omega} (\omega f) \times \overline{(\omega g)}$$

Now Ω^* is a subset of $\mathbb{C}^\Omega$, and, by the orthogonality of irreducible characters (see, for example, [9], Theorem V.5.8), Ω^* is an orthogonal basis for $\mathbb{C}^\Omega$. If we define subspaces S_x', T_x' of $\mathbb{C}^\Omega$ analogous to S_x, T_x, a dimension argument readily gives subsets of Ω^* as bases of S_x' and T_x'.

LEMMA 4. For $x \in \Omega^*$ let

$$S_x' = \{f \in \mathbb{C}^\Omega: \text{ there exists } f' \in \mathbb{C}^{\mathbb{C}} \text{ such that } f = xf'\}$$

and

$$T_x' = S_x' - \langle S_y': y \in \langle x \rangle, \langle y \rangle \neq \langle x \rangle \rangle$$

Then $\langle x \rangle$ is an orthogonal basis for S_x' and $\{y: \langle y \rangle = \langle x \rangle\}$ is an orthogonal basis for T_x'.

THEOREM 1. In the notation established in §4,

(a) the dimension of T_x is $\phi(o(x))$, where $o(x)$ is the order of x and ϕ is Euler's ϕ-function

(b) if $T_x \neq T_y$ then T_x and T_y are orthogonal

(c) $\mathbb{R}^\Omega$ is the orthogonal direct sum of the distinct T_x

(d) for every subset I of $\{1,\ldots,n\}$, $V_I = \oplus\{T_x: x \in \prod_{j \in I} \langle \chi_j \rangle\}$

(e) $T_x \subseteq W_{I(x)}$, where $I(x) = \{j: \omega_j x \neq 1\}$

Proof. The orthogonal basis found for T_x' in Lemma 4 is invariant under complex conjugation, because $\bar{y} = y^{-1}$. From this an orthogonal basis for T_x may be found by replacing $\{y,\bar{y}\}$ by $\{y + \bar{y}, (y - \bar{y})i\}$ if $y \neq \bar{y}$, and $\{y\}$ by $\{y\}$ if $y = \bar{y}$. Thus $\dim(T_x) = \dim(T_x') =$ the number of generators of $\langle x \rangle = \phi(o(x))$. Moreover, Lemma 4 shows that if $\langle x \rangle \neq \langle y \rangle$ then T_x' is orthogonal to T_y', so T_x is orthogonal to T_y. Thus (a) and (b) are proven. Since $\dim(T_x)$ is equal to the number of elements y with $T_x = T_y$, a counting argument gives (c). Replacing $\{1,\ldots,n\}$ by I in (c) gives (d).

To prove (e), note that $I(x)$ is the smallest subset of $\{1,\ldots,n\}$ such that $S_x \subseteq V_{I(x)}$. But (d) can be rewritten as $V_J = \oplus\{T_y: I(y) \subseteq J\}$. If $z \in \langle y \rangle$ then $I(z) \subseteq I(y)$, so $T_y \subseteq V_J$ if and only if $S_y \subseteq V_J$. Thus, if J is a proper subset of $I(x)$, V_J is a direct sum of spaces T_y different from, and hence orthogonal to, T_x. Therefore $T_x \subseteq V_{I(x)}$ but T_x is orthogonal to V_J for all proper subsets J of $I(x)$. It follows that $T(x) \subseteq W_{I(x)}$.

6. APPLICATION TO BLOCKING

If H is a subgroup of Ω^* then Ann H, the annihilator of H, is defined to be

$$\bigcap_{x \in H} Ann(x) = \{\omega \in \Omega: \omega x = 1 \text{ for all } x \text{ in } H\}$$

Annihilators of subgroups of Ω are similarly defined. It can be shown (see [9], Section V.6) that these two operations of taking annihilators are mutually inverse bijections on the sets of subgroups of Ω and of Ω^*, and moreover $|Ann\ H| = |\Omega^*|/|H|$.

THEOREM 2. Let H be a subgroup of Ω^*. If the blocks of treatments are the cosets of Ann H, then $V_{BLOCKS} = \oplus\{T_x: x \in H\}$. In particular, if $x \notin H$ then T_x is orthogonal to V_{BLOCKS}.

Proof.   If $x \in H$ then $Ann(x) \supseteq Ann\ H$. Thus elements of $\mathbb{R}^\Omega$ which are constant on each coset of $Ann(x)$ are also constant on each coset of $Ann(H)$. By the remarks after the definition of S_x, this shows that $S_x \subseteq V_{BLOCKS}$. Now $T_x \subseteq S_x$, so $T_x \subseteq V_{BLOCKS}$. Hence $V_{BLOCKS} \supseteq \oplus\{T_x: x \in H\}$. However, there are $|\Omega^*: Ann\ H|$ blocks, so $dim(V_{BLOCKS}) = |\Omega^*: Ann\ H| = |H|$, and by Theorem 1, $dim(T_x)$ is equal to the number of y such that $T_x = T_y$, so $|H| = dim(\oplus\{T_x: x \in H\})$. The result now follows.

COROLLARY. Let G be a subgroup of Ω. If the blocks of treatments are the cosets of G then $V_{BLOCKS} = \oplus\{T_x: x \in Ann\ G\}$.

Usually the experimenter specifies the block size b and some subsets J of $\{1,\dots,n\}$ such that W_J must not be confounded with blocks. Then, by Theorem 2, the problem for the algebraist is to find a subgroup H of Ω^* such that $|\Omega^*: H| = b$ and for all $x \in H$, $I(x)$ is not one of the specified subsets J. Although a solution is not always possible, the algebraic problem is generally more amenable to a quick solution, or quick decision as to impossibility, than the original problem. On the other hand the experimenter may sometimes propose a blocking scheme according to the cosets of some subgroup G of Ω. Then the Corollary to Theorem 2 can be used to see if the confounding is suitable.

REMARK. In the blocking schemes described in Theorem 2, T_x is confounded if and only if S_x is confounded. Thus one might question the need to define the spaces T_x. There are two important reasons why we do this. The first is that we require a decomposition of $\mathbb{R}^\Omega$ into disjoint subspaces, which the S-spaces do not provide. The second is that, for many x, $S_x - S_1 \not\subseteq W_{I(x)}$. A busy experimenter, if told that S_x is confounded, may check that $W_{I(x)}$ is of no importance and forget to check $W_{I(y)}$ for $y \in \langle x \rangle$, $y \neq 1$. Thus it is important to list the component spaces T_y. The only possible exception to this is when $I(y) = I(x)$ for all $y \in \langle x \rangle$, $y \neq 1$, which happens, for example, when $x = \chi_i$ and m_i is not prime.

At the other extreme, the proof of Theorem 1 suggests that the spaces T_x could be further broken down into orthogonal subspaces of dimension 1 or 2. If $o(x) \neq 2$, these new subspaces have bases $\{y + \bar{y}, (y - \bar{y})i\}$ if $\langle y \rangle = \langle x \rangle$, $y \neq \bar{y}$. (If $\langle y \rangle = \langle x \rangle$ and $y = \bar{y}$ then $o(x) = 2$ so T_x has dimension 1 already.) However, in the absence of a more sophisticated blocking scheme than that described in Theorem 2, there appears to be no advantage to this finer decomposition.

7. HISTORICAL AND RELATED NOTES

The special case of Theorems 1 and 2 in which Ω is an elementary abelian p-group for some prime p has been known and used for over 30 years. The theory was developed in [6, 7, 5, 10]. An equivalent theory, in terms of projective geometry over GF(p), was given in [4] and [3].

The great simplification when Ω is an elementary abelian group is that $T_x = S_x - V_\emptyset$ for all $x \neq 1$. Since $V_\emptyset$ is always removed from the space $\mathbb{R}^\Omega$ in statistical work (this accounts for the well-known rule-of-thumb "degrees of freedom are one less than you think"), no great distinction needs to be drawn between S_x and T_x. Thus the problems of having overlapping but unequal S_x, and of having S_x not contained in $W_{I(x)}$, do not arise. Familiarity with the elementary abelian case did not prepare people for the possibility of these problems. For example, in [11], where some steps towards the results in this paper are taken, a certain puzzlement was expressed at the fact that if S_x is confounded so also are some other $S_y \neq S_x$, and for such y we may not have $I(y) = I(x)$.

The results in this paper were presented in statistical language, with some examples, in [2]. To avoid reference to character theory, the authors used the bilinear form defined in Equation (4.2). Generalizations to more complicated blocking patterns are in [1].

8. ACKNOWLEDGEMENTS

The author is grateful to H. D. Patterson, N. J. Young and L. G. Kovács for helpful suggestions.

REFERENCES

1. R. A. Bailey, Patterns of confounding in factorial designs, Biometrika 64 (1977), 597–603.

2. R. A. Bailey, F. H. L. Gilchrist and H. D. Patterson, Identification of effects and confounding patterns in factorial designs, Biometrika 64 (1977), 347–354.

3. R. C. Bose, Mathematical theory of the symmetrical factorial design, Sankhyā 8 (1947), 107–166.

4. R. C. Bose and K. Kishen, On the problem of confounding in general symmetrical factorial design, Sankhyā 5 (1940), 21–36.

5. D. J. Finney, The fractional replication of factorial arrangements, Ann. Eugen. 12 (1945), 291–301.

6. R. A. Fisher, The theory of confounding in factorial experiments in relation to the theory of groups, Ann. Eugen. 11 (1942), 341–353.

7. R. A. Fisher, A system of confounding for factors with more than two alternatives, giving completely orthogonal cubes and higher powers, Ann. Eugen. 12 (1945), 283–290.

8. R. A. Fisher, The design of experiments, Oliver and Boyd, 1966.

9. B. Huppert, Endliche Gruppen I, Springer-Verlag, 1967.

10. O. Kempthorne, A simple approach to confounding and fractional replication in factorial experiments, Biometrika 34 (1947), 255–272.

11. D. White and R. A. Hultquist, Construction of confounding plans for mixed factorial designs, Ann. Math. Statist. 36 (1965), 1256–71.

ON A CLASS OF ASSOCIATION SCHEMES DERIVED
FROM LATTICES OF EQUIVALENCE RELATIONS

T. P. SPEED

Department of Mathematics
University of Western Australia
Nedlands, Western Australia
Australia

R. A. BAILEY*

Department of Mathematics
The Open University
Milton Keynes
England

1. INTRODUCTION

The results reported in this paper stemmed from the need to diagonalise some
families of real symmetric matrices which arose in the *analysis of variance*
[13, 17]. These families constitute the *association algebras* [15], or *adja-
cency algebras* [11], of certain *association schemes*, but they have a particular structure which permits direct arguments to be used in place of the
usual methods based on the regular representation of the algebra. Our schemes
are all related to a finite modular lattice of equivalence relations on the
base set and it turns out that the desired diagonalisation can be carried
out rather easily by exploiting this relation and making use of the Möbius
function of the lattice. In order to motivate and explain our procedure, we
devote the remainder of this introduction to a discussion of a simple yet
non-trivial example of one such scheme.

Current affiliation: Statistics Department, Rothamsted Experimental Station,
Harpenden, England.

Let X = {1,2,...,m} × {1,2,...,n} denote the set of ordered pairs (i,j), or just ij when no confusion is possible, $1 \leq i \leq m$, $1 \leq j \leq n$, and let us consider the family of all real symmetric matrices $\Gamma = (\gamma_{iji'j'})$ defined over X (indexed lexicographically) which for all i,j,k,ℓ, i' $\neq$ i, j' $\neq$ j, k' $\neq$ k, ℓ' $\neq$ ℓ, $1 \leq i,i',k,k' \leq m$, $1 \leq j,j',\ell,\ell' \leq n$ satisfy the following

Constraints	Common entry for pairs
$\gamma_{ijij} = \gamma_{k\ell k\ell}$	on the diagonal
$\gamma_{ijij'} = \gamma_{k\ell k\ell'}$	with common first, different second component
$\gamma_{iji'j} = \gamma_{k\ell k'\ell}$	with common second, different first component
$\gamma_{iji'j'} = \gamma_{k\ell k'\ell'}$	with no common component

For m = 3, n = 2 such matrices may be written in the form

$$\Gamma = \begin{bmatrix} \alpha & \beta & \gamma & \delta & \gamma & \delta \\ \beta & \alpha & \delta & \gamma & \delta & \gamma \\ \gamma & \delta & \alpha & \beta & \gamma & \delta \\ \delta & \gamma & \beta & \alpha & \delta & \gamma \\ \gamma & \delta & \gamma & \delta & \alpha & \beta \\ \delta & \gamma & \delta & \gamma & \beta & \alpha \end{bmatrix}$$

$$= \alpha A_e + \beta A_r + \gamma A_c + \delta A_u$$

where

$$A_e = \begin{bmatrix} 1 & 0 & 0 & 0 & 0 & 0 \\ 0 & 1 & 0 & 0 & 0 & 0 \\ 0 & 0 & 1 & 0 & 0 & 0 \\ 0 & 0 & 0 & 1 & 0 & 0 \\ 0 & 0 & 0 & 0 & 1 & 0 \\ 0 & 0 & 0 & 0 & 0 & 1 \end{bmatrix} \qquad A_r = \begin{bmatrix} 0 & 1 & 0 & 0 & 0 & 0 \\ 1 & 0 & 0 & 0 & 0 & 0 \\ 0 & 0 & 0 & 1 & 0 & 0 \\ 0 & 0 & 1 & 0 & 0 & 0 \\ 0 & 0 & 0 & 0 & 0 & 1 \\ 0 & 0 & 0 & 0 & 1 & 0 \end{bmatrix}$$

$$A_c = \begin{bmatrix} 0 & 0 & 1 & 0 & 1 & 0 \\ 0 & 0 & 0 & 1 & 0 & 1 \\ 1 & 0 & 0 & 0 & 1 & 0 \\ 0 & 1 & 0 & 0 & 0 & 1 \\ 1 & 0 & 1 & 0 & 0 & 0 \\ 0 & 1 & 0 & 1 & 0 & 0 \end{bmatrix} \qquad A_u = \begin{bmatrix} 0 & 0 & 0 & 1 & 0 & 1 \\ 0 & 0 & 1 & 0 & 1 & 0 \\ 0 & 1 & 0 & 0 & 0 & 1 \\ 1 & 0 & 0 & 0 & 1 & 0 \\ 0 & 1 & 0 & 1 & 0 & 0 \\ 1 & 0 & 1 & 0 & 0 & 0 \end{bmatrix}$$

The subscripts on these *association* matrices are *e* for *equality*, *r* for *row* association, *c* for *column* association, and *u* for the association of *unrelated* pairs, this description referring to the representation of X as the cells in Figure 1.

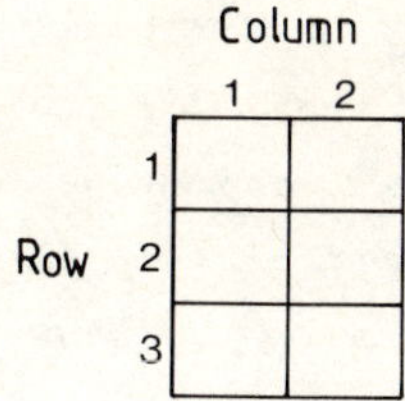

Figure 1

It is clear that we can diagonalise all such Γ when and only when we can diagonalise A_e, A_r, A_c and A_u, and we now turn to a lattice of equivalences on X which helps us do just this.

We begin by observing that the relations θ_e, θ_r, θ_c and θ_a defined on X through the relationship matrices (their characteristic functions) $R_e = A_e$, $R_r = A_e + A_r$, $R_c = A_e + A_c$ and $R_a = A_e + A_r + A_c + A_u$ respectively are all *equivalence relations* which have *equi-potent* equivalence classes. Let us call them uniform equivalences. Furthermore, if we denote by $\circ$ and & the binary operations of relation *composition* and *conjunction*, then it is easily checked that

$$\theta_r \circ \theta_c = \theta_c \circ \theta_r = \theta_a, \quad \theta_r \, \& \, \theta_c = \theta_c \, \& \, \theta_r = \theta_e$$

Indeed we find that these four equivalence relations form a distributive lattice (with the reverse of the usual partial order of relations) having $\circ$ as inf and & as sup, see Figure 2(i).

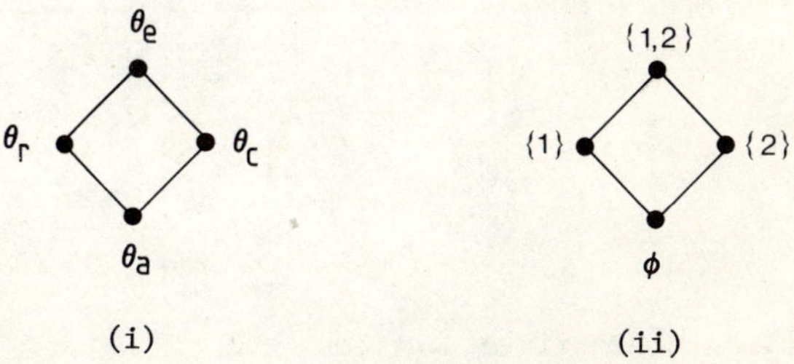

Figure 2

A further simplification occurs when we display this lattice as a lattice of subsets of the set of (two) indices labelling elements of X, see Figure 2(ii). For in this view $\theta_r = \theta_{\{1\}}$ is the equivalence defined by $(x_1,x_2) \equiv (x_1',x_2')\theta_{\{1\}}$ iff $x_1 = x_1'$, i.e. by having a *common first coordinate*, with a similar definition for $\theta_c = \theta_{\{2\}}$, $\theta_a = \theta_\phi$ and $\theta_e = \theta_{\{1,2\}}$ (identity).

The next step is to show that the introduction of these equivalence relations helps us solve the original diagonalisation problem. It is easy to check in this special case, and is a consequence of the lattice closure and uniformity of our set of equivalences in general, that the matrices R_e, R_r, R_c and R_a can be *scaled* so as to become a *commuting* family of *symmetric idempotent* matrices. We denote these scaled matrices by P with the same subscripts and they are $P_e = R_e$, $P_r = (1/n)R_r$, $P_c = (1/m)R_c$ and $P_a = (1/mn)R_a$. The commuting properties are all consequences of the fact that P_e is the identity matrix and $P_r P_c = P_c P_r = P_a$.

Our final task is to *orthogonalise* these commuting symmetric idempotents and this is most easily done by making use of their natural ordering: $U \le V$ iff $UV = VU = U$, equivalently, iff $R(U) \subseteq R(V)$, where $R(U)$ denotes the range of the matrix (operator) U. For if $U \le V$ in this sense, then it is easy to check that U and $V - U$ are orthogonal commuting symmetric idempotents with $V - U \le V$, and this process may be continued until minimal non-zero elements are obtained. They will all be orthogonal and our end will have been achieved. In our current example the partially-ordered set (abbrev. poset) turns out to be the lattice of Figure 2 and the orthogonalised idempotents are thus $S_a = P_a$, $S_r = P_r - P_a$, $S_c = P_c - P_a$ and $S_e = P_e - P_r - P_c + P_a$.

It is now apparent that we have obtained a set of pairwise orthogonal symmetric idempotents which can be expressed as *linear* combinations of our original association matrices, for each of the steps $A \to R$, $R \to P$ and $P \to S$ involved a linear expression. The result of this procedure may be written in matrix form as follows

$$
\begin{bmatrix} S_a \\ S_c \\ S_r \\ S_e \end{bmatrix} = \frac{1}{mn} \begin{bmatrix} 1 & n-1 & m-1 & (m-1)(n-1) \\ 1 & -1 & m-1 & -(m-1) \\ 1 & n-1 & -1 & -(n-1) \\ 1 & -1 & -1 & 1 \end{bmatrix}^T \begin{bmatrix} A_e \\ A_r \\ A_c \\ A_u \end{bmatrix}
$$

and the inverse of this matrix of coefficients is just the same matrix with the multiplier $(mn)^{-1}$ omitted. Thus we have obtained the *spectral decomposition* of our association matrices and the solution to our original problem.

Having discussed our example in some detail, we can now briefly explain the contents of this paper. In §2 below we carry out the final step above under quite general assumptions, going from certain systems of equivalence relations (R) via projections (P) to orthogonalised projections (S). Having done this we can then show the connection between such systems of equivalence relations and association schemes. Following on from this we describe in §3 a class of such systems of equivalence relations obtained from finite distributive lattices, and the application of our earlier results to this class provides the solution to many of the diagonalisation problems which arise in the analysis of variance.

2. LATTICES OF COMMUTING UNIFORM EQUIVALENCES

In this section we will denote equivalence relations on a finite set X by θ, ψ etc., writing $\theta(x) = \{y \in X: (x,y) \in \theta\}$ and R_θ for the binary matrix on X which is the characteristic function of θ. We say that an equivalence relation θ is *uniform* if the cardinality $|\theta(x)|$ does not vary with $x \in X$ and we offer the following simple reformulation of this notion.

LEMMA 1. The equivalence relation θ is uniform if and only if $R_\theta^2 = mR_0$ for some natural number m. In this case $m = |\theta(x)|$ for all $x \in X$ and $P_\theta = (1/m)R_\theta$ is a symmetric idempotent matrix over X.

For a uniform equivalence relation θ this integer $m = m_\theta$ will be termed the *block size* of θ.

Let us write the *conjunction* of the equivalences θ,ψ by $\theta \& \psi$, this being defined by $(x,y) \in \theta \& \psi$ iff $(x,y) \in \theta$ and $(x,y) \in \psi$. The relational *composition* or *product* $\theta \circ \psi$ is the (not necessarily equivalence) relation: $(x,y) \in \theta \circ \psi$ iff there exists $z \in X$ such that $(x,z) \in \theta$ and $(z,y) \in \psi$, i.e. iff $\theta(x) \cap \psi(y)$ is non-empty. We say that θ and ψ *commute* (the term *permute* is also used) if $\theta \circ \psi = \psi \circ \theta$; it is well known and easy to prove that $\theta \circ \psi$ is an equivalence relation if θ commutes with ψ. With these notions we can prove:

LEMMA 2. Let θ and ψ be two commuting uniform equivalence relations with block sizes m_θ, m_ψ respectively, and suppose that their conjunction $\theta \& \psi$ is

also uniform with block size $m_{\theta\&\psi}$. Then $R_\theta R_\psi = R_\psi R_\theta$ and $\theta \circ \psi = \psi \circ \theta$ is a
uniform equivalence with block size

$$m_{\theta\circ\psi} = \frac{m_\theta m_\psi}{m_{\theta\&\psi}}$$

Proof. $R_\theta R_\psi(x,y) = |\theta(x) \cap \psi(y)|$ and by our assumptions this is either
zero when $(x,y) \notin \theta \circ \psi$, or a natural number $m = m_{\theta\&\psi} \neq 0$ when $(x,y) \in \theta \circ \psi$.
But then the assumption $\theta \circ \psi = \psi \circ \theta$ implies that $R_\psi R_\theta(x,y)$ is 0 or m under
the same circumstances, i.e. $R_\theta R_\psi = R_\psi R_\theta$. The diagram of Figure 3 shows us
what is happening, although there we can only depict X as a single $\theta \circ \psi$-
block. Now we have seen that $R_{\theta\circ\psi} = m^{-1}R_\theta R_\psi = m^{-1}R_\psi R_\theta$ where $m = m_{\theta\&\psi}$, and so

$$R_{\theta\circ\psi}^2 = m^{-2}R_\theta R_\psi R_\theta R_\psi = m^{-2}R_\theta^2 R_\psi^2 = m^{-2}m_\theta m_\psi R_\theta R_\psi = m^{-1}m_\theta m_\psi R_{\theta\circ\psi}$$

It follows that (under the hypotheses of Lemma 2) $P_\theta P_\psi = P_{\psi\circ\theta} = P_\psi P_\theta$.

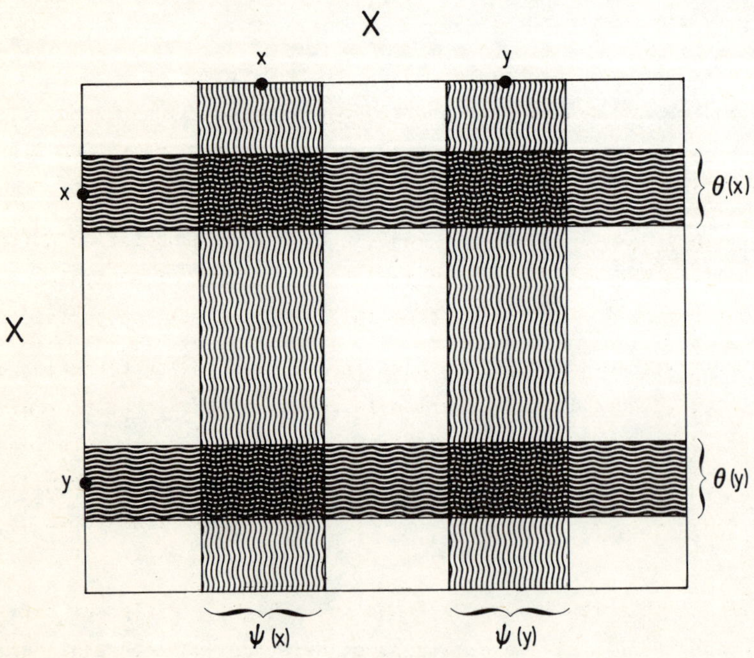

Figure 3

We now move towards our first main result, a description of the minimal orthogonal symmetric idempotents of the $\mathbb{R}$-algebra generated by the matrices $\{R_\theta : \theta \in L\}$ of a finite lattice L of commuting uniform equivalence relations. This description makes use of the *Möbius function* $\mu: L \times L \to \mathbb{Z}$ of L, see [4,1]. For our present purposes it suffices to define μ as the *inverse* of the *zeta function* $\zeta: L \times L \to \mathbb{Z}$ of L, where $\zeta(\theta,\psi) = 1$ if $\theta \leq \psi$, and equals 0 otherwise, and "inverse" is in the sense of matrices over L; that is

$$\sum_{\psi \in L} \mu(\theta,\psi)\zeta(\psi,\chi) = \delta(\theta,\chi)$$

where $\delta(\theta,\chi)$ denotes the Kronecker delta function over $L \times L$.

PROPOSITION 1. Let L be a finite family of commuting uniform equivalence relations on a finite set X which is closed under the operations of conjunction (&) and composition (∘). Then the $\mathbb{R}$-algebra generated by the relationship matrices $\{R_\theta : \theta \in L\}$ is commutative and its minimal orthogonal symmetric idempotents are the matrices

$$S_\theta = \sum_{\psi \leq \theta} \mu(\psi,\theta)P_\psi = \sum_\psi \mu(\psi,\theta)P_\psi \tag{1}$$

where P_ψ is the symmetric idempotent obtained by scaling R_ψ, and μ is the Möbius function of the lattice (L;&,∘) with join (sup) operation & and meet (inf) operation ∘.

REMARK 1. The partial order on L induced by the join operation & (resp. meet operation ∘) is the *reverse* of the inclusion ordering of equivalence relations. Our reason for deviation from this practice will be explained below.

REMARK 2. It will usually be the case in our examples that the largest (resp. smallest) equivalence in our ordering is *equality* (resp. the *trivial* relation with a single block) and so the $\mathbb{R}$-algebra generated by the relationship matrices will then contain the identity I (resp. the all-1 matrix J). However the arguments we give do not require these assumptions so we do not make them.

Proof. This result can be found in a slightly different form in [1] IV.4. However we can give a direct proof here for completeness, using only the fact that μ is the inverse of ζ. This means that the defining relations (1) are equivalent to

$$P_\theta = \sum_{\psi \leq \theta} S_\psi = \sum_\psi \zeta(\psi,\theta) S_\psi \qquad\qquad (2)$$

Our proof uses the equality $S_\theta P_\chi = \zeta(\theta,\chi) S_\theta$, $\theta,\chi \in L$, which we establish by induction on the position of θ in the lattice. If θ is the least element 0 of the lattice, then $S_0 = P_0$ whence $S_0 P_\chi = P_0 P_\chi = P_{0 \circ \chi}$ (by the remark following Lemma 2) and this equals P_0. Now suppose $\theta \neq 0$ is an arbitrary element of the lattice and that $S_\omega P_\chi = \zeta(\omega,\chi) S_\omega$ holds for all $\omega < \theta$. We use (2) to write

$$S_\theta = P_\theta + \sum_\omega [\delta(\omega,\theta) - \zeta(\omega,\theta)] S_\omega$$

Now we observe that $\delta(\omega,\theta) - \zeta(\omega,\theta) \neq 0$ only when $\omega < \theta$ and for such ω our inductive hypothesis holds.   Thus

$$S_\theta P_\chi = P_\theta P_\chi + \sum_\omega [\delta(\omega,\theta) - \zeta(\omega,\theta)] S_\omega P_\chi$$

$$= P_{\theta \circ \chi} + \sum_\omega [\delta(\omega,\theta) - \zeta(\omega,\theta)] \zeta(\omega,\chi) S_\omega$$

$$= P_{\theta \circ \chi} + \zeta(\theta,\chi) S_\theta - \sum_\omega \zeta(\omega,\theta) \zeta(\omega,\chi) S_\omega$$

which equals $\zeta(\theta,\chi) S_\theta$ by (2) since $\zeta(\omega,\theta) \zeta(\omega,\chi) = \zeta(\omega,\theta \circ \chi)$. This completes the induction step and establishes the desired equality.

An easy calculation now shows that the $\{S_\theta\}$ are pairwise orthogonal idempotents. For using the result just obtained

$$S_\omega S_\theta = S_\omega \sum_\psi \mu(\psi,\theta) P_\psi = \sum_\psi \mu(\psi,\theta) \zeta(\omega,\psi) S_\omega = \delta(\omega,\theta) S_\omega$$

To see that the $\{S_\theta\}$ are minimal idempotents in the $\mathbb{R}$-algebra generated by $\{R_\theta\}_{\theta \in L}$ we use their orthogonality and (2), whilst the commutativity of this algebra is a consequence of Lemma 2.

This proposition clearly generalises the last part of the procedure described in our introduction and it only remains to make the connection with association schemes. Once again this involves the Möbius function of L.

Take a family L of commuting uniform equivalences on a set X which is a
lattice under the operations & and ∘. For each $\theta \in L$ we define another relation (an *association*!) $\bar{\theta}$ which is *not* an equivalence relation in general
by writing

$$(x,y) \in \bar{\theta} \text{ iff } (x,y) \in \theta \text{ and, for all } \psi \geq \theta, \ (x,y) \notin \psi \tag{3}$$

If we write A_θ for the binary matrix over X which is the characteristic
function of the association $\bar{\theta}$, then (3) is clearly equivalent to

$$R_\theta = \sum_{\psi \geq \theta} A_\psi = \sum_\psi \zeta(\theta,\psi) A_\psi \tag{4}$$

By Möbius inversion this is equivalent to writing

$$A_\theta = \sum_\psi \mu(\theta,\psi) R_\psi \tag{5}$$

The matrices we get by the construction just described define an *association scheme* [15, 16], provided that the family L does contain the identity
and the trivial equivalence relation, and this is our next result.

PROPOSITION 2. Let L be a finite family of commuting uniform equivalence
relations on a set X which contains the identity and the trivial equivalence
relations, and is closed under the operations of conjunction and composition.
Then the matrices $\{A_\theta : \theta \in L\}$ defined by (5) above, where μ is the Möbius
function of the lattice $(L; \&, \circ)$ and R_θ the relationship matrix of the equivalence $\theta \in L$, define an association scheme.

Proof.   It is necessary to show that there exist real constants $\{a_{\theta\psi\chi}\}$
such that

$$A_\theta A_\psi = \sum_\chi a_{\theta\psi\chi} A_\chi \tag{6}$$

and this is an easy consequence of Proposition 1 and equations (1), (2) and
(4) above. For $A_\theta A_\psi$ can be written as a linear combination of the $\{R_\chi\}$ and
hence, via the $\{P_\chi\}$, of the $\{S_\chi\}$. But by reversing this argument, the $\{S_\chi\}$
can be written as linear combinations of the $\{A_\chi\}$ and so (6) follows. If
the largest and smallest equivalences are denoted by ι and 0 respectively,
then $A_\iota = I$, the identity matrix, and $R_0 = J = \sum_{\theta \in L} A_\theta$, an all-1 matrix over
X.

The two-stage procedure which we saw in our introductory example is the application of Propositions 2 and 1 in that order, passing from $\{A_\theta\}$ to $\{S_\theta\}$ via $\{R_\theta\}$ and $\{P_\theta\}$. If we write the resulting linear equations

$$S_\psi = \frac{1}{|X|} \sum_\theta q_{\theta\psi} A_\theta$$

with inverse, apart from the multiplier $|X|$,

$$A_\psi = \sum_\theta p_{\theta\psi} S_\theta$$

then various expressions for the elements of the matrices $P = (p_{\theta\psi})$ and $Q = (q_{\theta\psi})$ can be given. We readily find that in terms of the Möbius function of L we have

$$p_{\theta\psi} = \sum_{\omega \geq \theta \& \psi} m_\omega \mu(\psi,\omega)$$

whilst

$$q_{\theta\psi} = \sum_{\omega \leq \theta \circ \psi} n_\omega \mu(\omega,\psi)$$

where $n_\omega = m_\omega^{-1} |X|$ is the number of ω-blocks. These expressions are only helpful if the lattice is not too large. However it is a straightforward task to derive *nesting and crossing rules* which are more useful for the inductively-defined class of *simple orthogonal block structures* [13] and we will give some details below. Note that if a symmetric matrix Γ can be written in the form $\sum_\psi \gamma_\psi A_\psi$ where the $\{A_\psi\}$ are related to a lattice of commuting uniform equivalences as above, then its spectral decomposition is $\sum \xi_\theta S_\theta$, where the eigenvalues ξ_θ are given by

$$\xi_\theta = \sum_\psi p_{\theta\psi} \gamma_\psi$$

One of the standard methods of deriving association schemes involves *Kronecker* or *tensor* products of association matrices and it is of interest to relate this approach to our constructions. Suppose that L_1 and L_2 are finite lattices of commuting uniform equivalences defined on sets X_1 and X_2 respectively. Then $L = L_1 \times L_2$ is a lattice of commuting uniform equivalences defined componentwise on $X = X_1 \times X_2$ and the association matrices obtained via Proposition 2 satisfy

$$A_{\theta,\psi} = A_\theta \otimes A_\psi, \quad \theta \in L_1, \quad \psi \in L_2$$

Furthermore, the orthogonalised idempotents given by Proposition 1 have the form

$$S_{\theta,\psi} = S_\theta \otimes S_\psi, \quad \theta \in L_1, \quad \psi \in L_2$$

whilst the matrices of coefficients connecting these two families of matrices can be given by the formulae

$$P = P_1 \otimes P_2, \quad Q = Q_1 \otimes Q_2$$

For these last identities it is convenient to order the minimal idempotent matrices in a manner reverse to that adopted for ordering the association matrices, the corresponding entities of the product system then being ordered lexicographically. This construction will be called *crossing* and the various results we have noted *crossing rules*. All of these assertions are easily checked and so the proofs will be omitted.

The other important construction in statistics is known as *nesting*, and although the full significance of it will not be evident until §3 below, we give some details now. As with crossing, we take lattices L_1 and L_2 of commuting uniform equivalences on sets X_1 and X_2, but this time we make use of their *ordinal sum* [8], here denoted by $L = L_1/L_2$, as a lattice of equivalence relations on $X = X_1 \times X_2$. The lattice L is obtained by placing L_2 above L_1, in the sense of posets, and identifying its least element with the greatest element of L_1. It arises as the lattice of the following equivalence relations on $X_1 \times X_2$: $(\theta, 0_2)$, $\theta \in L_1$ but $\theta \neq \iota_1$, where ι_1 is the identity relation over X_1 and 0_2 the trivial relation over X_2, and (ι_1, ψ), $\psi \in L_2$. Proceeding as in Proposition 2 we obtain an association scheme with matrices

$$A_{\theta,-} = A_\theta \otimes J_2, \quad \theta \neq \iota_1, \quad \theta \in L_1$$

$$A_{1,\psi} = I_1 \otimes A_\psi, \quad \psi \in L_2$$

where I_1 and J_2 denote the identity and all-1s matrix over X_1 and X_2 respectively. The orthogonalised idempotents given by Proposition 1 turn out to be

$$S_{\theta,0} = S_\theta \otimes S_0, \quad \theta \in L_1, \quad 0 = 0_2$$

$$S_{+,\psi} = I_1 \otimes S_\psi, \quad \psi \in L_2, \quad \psi \neq 0_2$$

More care is needed in the formulation of a nesting rule for the P and Q matrices of coefficients connecting the association matrices with the orthogonalised idempotents. For convenient expressions we must begin listing associations with the identity ι and end with the trivial association 0, i.e. go down the lattice, whilst for the idempotents we do this in the reverse order. With these conventions a similar ordering arises for the P and Q matrix of L_1/L_2:

$$
P_{1/2} = \begin{array}{cc} L_2 \text{ (begin at } \iota_2) & L_1 \text{ (without } \iota_1) \\ \left[\begin{array}{c|c} \text{All rows equal to first row of } P_2 & |X_2|P_1 \text{ without its first column} \\ \hline P_2 \text{ without its first row} & 0 \end{array}\right] & \begin{array}{l} L_1 \text{ (begin at } 0_1) \\[1em] L_2 \text{ (without } 0_2) \end{array} \end{array}
$$

$$
Q_{1/2} = \begin{array}{cc} L_1 \text{ (begin at } 0_1) & L_2 \text{ (without } 0_2) \\ \left[\begin{array}{c|c} \text{All rows equal to first row of } Q_1 & |X_1|Q_2 \text{ without its first column} \\ \hline Q_1 \text{ without its first row} & 0 \end{array}\right] & \begin{array}{l} L_2 \text{ (begin at } \iota_2) \\[1em] L_1 \text{ (without } \iota_1) \end{array} \end{array}
$$

The proofs of these *nesting rules* are also straightforward and so omitted.

We remark here that many of the naturally arising examples of such schemes can be related to a group action on X, see §3 below, in which case the algebra we are analysing is the centraliser algebra of a generally transitive permutation representation of the group in question, see [3]. In view of this it is of interest to close this section with an example which is *not* derived from a permutation representation.

EXAMPLE. Let Λ be the 9×9 Latin square which is the multiplication table of the elementary abelian group of order 9, and let us view it as a set X of 81 ordered triples (i,j,k), $1 \le i,j,k \le 9$, where the first two

indices correspond to row and column labels, and the third index labels the
Latin letters, only those combinations which arise in the definition of the
square belonging to X.

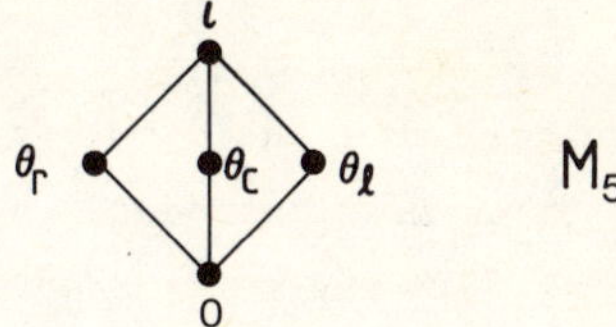

Figure 4

For our equivalence relations we have θ_r, θ_c and θ_ℓ corresponding to equality
of row, column and Latin index respectively, and to these we add the identity
ι and the trivial equivalence relation, here denoted by 0. It is easy to see
that these equivalence relations are uniform, that they commute, and that they
give the 5-element modular non-distributive lattice M_5 under the operations
of conjunction and composition, see Figure 4.

The association scheme derived from this family of equivalences is *not*
what is known as a *Latin-square type* in the literature [16] but has doubtless
occurred to many people. Our interest here is not in its existence or construction, but in its relation to the lattice of equivalences. The association matrices are $A_\iota = R_\iota = I$, the identity matrix over X, $A_r = R_r - R_\iota$,
$A_c = R_c - R_\iota$, $A_\ell = R_\ell - R_\iota$ and $A_u = R_0 - R_r - R_c - R_\ell + 2R_\iota$ where $R_0 = J$,
the all-1 matrix over X. Similarly the minimal idempotents are

$$S_0 = \frac{1}{81} R_0$$

$$S_r = \frac{1}{9} R_r - \frac{1}{81} R_0$$

$$S_c = \frac{1}{9} R_c - \frac{1}{81} R_0$$

$$S_\ell = \frac{1}{9} R_\ell - \frac{1}{81} R_0$$

$$S_\iota = I - \frac{1}{9}(R_r + R_c + R_\ell) + \frac{2}{81} R_0$$

All of these relations are consistent with the values taken by the Möbius
function of M_5 which are 1 or -1 except for $\mu(0,\iota) = 2$.

Although we do not give any details here, it can be shown that this
scheme does not derive from a group action on X, see [2].

3. REPRESENTING LATTICES AS LATTICES OF COMMUTING UNIFORM EQUIVALENCES

Our original aim was to carry out the decomposition described in §2 above
for the class of *block structures* introduced by Throckmorton [18] as *complete
balanced* block structures. In the terminology of this paper, these structures
can be described as association schemes derived from finite distributive
lattices of commuting uniform equivalences in the manner detailed in Proposition 2, and our main task of this section is to show that *every* finite
distributive lattice can arise as a lattice of commuting uniform equivalences.
Such a result links our work with that of Jónsson [12], see also [5] and [6],
where so-called Type I representations of lattices are discussed. One result
from this lattice theory is the following: any lattice of commuting (also
known as *permutable*) equivalence relations is necessarily modular; indeed it
satisfies the stronger Arguesian axiom, a self-dual axiom implying modularity
which is related to the validity of Desargues' axiom in the associated projective geometry. We will comment on these more general lattices later, and
for the moment content ourselves with the simpler task of representing finite
distributive lattices.

Let $(\pi, \leq)$ be a *finite partially ordered* set, abbreviated as *poset*, and
let us call a subset $a \subseteq \pi$ *hereditary* if $t \in a$ whenever $t \leq s$ for some $s \in a$.
Then the family $L(\pi)$ of all hereditary subsets of π is easily seen to be a
distributive lattice of subsets of π with largest element π and least element
ϕ. [Conversely, if L is any finite *distributive* lattice, then L has a non-
empty set $\pi(L)$ of join-irreducibles and the map from $a \in L$ to the hereditary
subset $\{t \in \pi(L): t \leq a\}$ of $\pi(L)$ defines an isomorphism of L with $L(\pi(L))$.
Thus we may suppose all finite distributive lattices are lattices of subsets
in this way, see [4] and [6].] Associate with every $t \in \pi$ a set X_t of $n_t \geq 2$
elements, and write $X = \Pi\{X_t: t \in \pi\}$, elements of X being written $\underline{x} = (x_t)_{t \in \pi}$.
For every hereditary subset $a \subseteq \pi$, i.e. for every $a \in L(\pi)$, define the relation θ_a on X as follows:

$$(\underline{x}, \underline{y}) \in \theta_a \text{ iff } \{t \in \pi: x_t = y_t\} \supseteq a$$

PROPOSITION 3.   The relations θ_a, $a \in L(\pi)$, are commuting uniform equivalences, and the set of all these is closed under conjunction (&) and composition ($\circ$). More precisely, we have θ_a & $\theta_b = \theta_{a \cup b}$, $\theta_a \circ \theta_b = \theta_b \circ \theta_a = \theta_{a \cap b}$ and $|\theta_a(\underline{x})| = \prod_{t \notin a} n_t$ for all $a,b \in L(\pi)$.

Proof.   These are all easy computations. Clearly if $(\underline{x},\underline{y}) \in \theta_a$ and $(\underline{x},\underline{y}) \in \theta_b$ then $\{t \in \pi: x_t = y_t\} \supseteq a \cup b$ and conversely, proving the first identity. Similarly if $(\underline{x},\underline{z}) \in \theta_a$ and $(\underline{z},\underline{y}) \in \theta_b$ then $\{t \in \pi: x_t = z_t\} \supseteq a$ and $\{t \in \pi: z_t = y_t\} \supseteq b$ whence $\{t \in \pi: x_t = y_t\} \supseteq a \cap b$, whilst if $\underline{x}$ and $\underline{y}$ agree on coordinates labelled by $a \cap b$, we may define $\underline{z} = (z_t)$ by $z_t = x_t$, $t \in a \backslash b$, $z_t = y_t$, $t \in b$ and arbitrarily off $a \cup b$. Such a $\underline{z}$ belongs to both $\theta_a(\underline{x})$ and $\theta_b(\underline{y})$ implying that $(\underline{x},\underline{y}) \in \theta_a \circ \theta_b$ and at the same time, that $\theta_a \circ \theta_b = \theta_{a \cap b} = \theta_b \circ \theta_a$. The cardinality of $\theta_a(\underline{x})$ is obvious.

COROLLARY.   If $L = L(\pi)$ is a finite distributive lattice represented as the lattice of all hereditary subsets of its set of all join-irreducibles, then the map $a \to \theta_a$ is a Type I representation of L.

Constructions analogous to the one just described can be given for some small non-distributive (necessarily Arguesian) lattices, but the absence of any useful structure theorem for these makes it difficult to carry out this task in any generality. For example, we can realise the lattice of Figure 5(i) as a lattice of commuting uniform equivalences using a pair of orthogonal Latin squares, but that of Figure 5(ii) already requires more effort. Part of the subgroup lattice of the elementary abelian group of order 16 works in this case but other examples seem much harder.

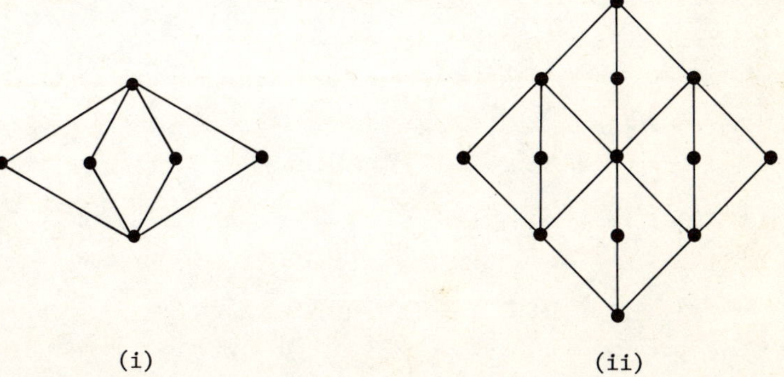

(i)   (ii)

Figure 5

Let us denote the lattice of equivalences of Proposition 3 above by $L = L(\pi, \{X_t : t \in \pi\})$. Our final proposition asserts that *all* distributive lattices of commuting uniform equivalences arise in this way. We continue to use the reverse of the usual ordering of equivalences, with lattice join as &, and meet as $\circ$.

PROPOSITION 4. Let L be a finite distributive lattice of commuting uniform equivalences on a set X and suppose that the largest (resp. smallest) element of L is the identity (resp. trivial) equivalence on X. Then there exists a poset π, a family $\{X_\tau : \tau \in \pi\}$ of sets indexed by π, a bijective mapping $f: X \to \Pi\{X_\tau : \tau \in \pi\}$ and a lattice isomorphism $g: L \to L(\pi, \{X_\tau : \tau \in \pi\})$ such that for all $\psi \in L$ and $x, y \in X$

$$(x,y) \in \psi \text{ iff } (f(x), f(y)) \in g(\psi) \tag{7}$$

REMARK. We are not assuming that X is finite.

Proof.   Clearly the poset π of this proposition must be $\pi = \pi(L)$, the poset of all join-irreducibles of L, and for $\psi \in L$ we write $a(\psi)$ equal to $\{\tau \in \pi(L) : \tau \leq \psi\}$ for the hereditary subset of π corresponding to ψ. The representation theorem for finite distributive lattices is embodied in the equations:

$$a(\psi \circ \chi) = a(\psi) \cap a(\chi), \quad a(0) = \phi$$
$$a(\psi \,\&\, \chi) = a(\psi) \cup a(\chi), \quad a(\iota) = \pi \tag{8}$$

We turn now to the definition of the sets X_τ, $\tau \in \pi$. If τ is a *minimal* element of π, equivalently, if τ is an *atom* of L, then we let X_τ be the set of τ blocks, i.e. $X_\tau = X/\tau$, and we let $x_\tau = \tau(x)$ denote the τ-block containing $x \in X$. For a *non-minimal* element τ of π we let τ_- denote the join of all elements of L strictly below τ, i.e.

$$\tau_- = \&\{\sigma \in L : \sigma < \tau\}$$

Since τ is join-irreducible, $\tau_- < \tau$. Now τ_- and τ are both uniform equivalences, and so each τ_--block is the (disjoint) union of the *same* number of τ-blocks, and in this case we let X_τ be any set of this cardinality. Finally, we take a bijection from the τ-blocks within a given τ_--block to the set X_τ and, for $x \in X$, let $x_\tau \in X_\tau$ denote the image under this bijection of $\tau(x)$.

Having now defined the sets $\{X_\tau : \tau \in \pi\}$ and for each $\tau \in \pi$ a map from X to X_τ, we define $f: X \to X = \Pi\{X_\tau : \tau \in \pi\}$ as follows:

$$f(x) = (x_\tau)_{\tau \in \pi}$$

Our main task is to prove that f is a bijection. Take an element $x^* = (x_\tau^*)_{\tau \in \pi} \in X$ and for each $\tau \in \pi$ let $X_\tau = \{x \in X : x_\tau = x_\tau^*\}$. We prove by induction on the height of $\psi \in L$ that

$$\text{for all } \psi \in L, \ \cap\{X_\tau : \tau \leq \psi\} \text{ is a } \psi\text{-block} \tag{9}$$

This is trivially true for $\psi = 0$ if we interpret an empty intersection as the set X, and true by definition for ψ a minimal join-irreducible, i.e. an atom of L. Now suppose that ψ is neither 0 nor an atom of L and that (9) is true for all $\chi < \psi$. We need to distinguish two cases. Firstly, if ψ is a (non-minimal) join-irreducible, then

$$\cap\{X_\tau : \tau \leq \psi\} = \cap\{X_\tau : \tau < \psi\} \cap X_\psi = \cap\{X_\tau : \tau \leq \psi_-\} \cap X_\psi$$

and since $\psi_- < \psi$, the assertion follows from the inductive hypothesis and the definition of x_ψ. Next, suppose that ψ is not join-irreducible but can be written $\psi = \kappa \vee \lambda = \kappa \ \& \ \lambda$ where $\kappa < \psi$, $\lambda < \psi$. Then we may write $\cap\{X_\tau : \tau \leq \psi\} = A \cap B \cap C$ where

$$A = \cap\{X_\tau : \tau \leq \kappa \circ \lambda\}, \ B = \cap\{X_\tau : \tau \leq \kappa, \tau \nleq \lambda\}, \ C = \cap\{X_\tau : \tau \leq \lambda, \tau \nleq \kappa\}$$

By the inductive hypothesis A is a $\kappa \circ \lambda$-block containing $A \cap B$, which is a κ-block, and $A \cap C$, which is a λ-block, and so (cf. Lemma 2) $(A \cap B) \cap (A \cap C)$ is a $\kappa \ \& \ \lambda$-block. This completes the inductive step and proves that $\cap\{X_\tau : \tau \in \pi\}$ is a ι-block, i.e. a singleton.

It follows from the preceding discussion that f is a bijection and our proof will be completed when the isomorphism g of the statement is exhibited and (7) proved. But this is easy: in the notation of Proposition 3 and that introduced prior to (8) above, we write $g(\psi) = \theta_{a(\psi)}$. The isomorphism is readily checked and (7) follows from (9) and the definition of the equivalences θ_a.

We now describe the simple connection between our poset-lattice construction and the nesting and crossing operations given on page 65. If π_1

and π_2 are two finite posets, we may write π_1/π_2 for their *poset ordinal sum*
[4] obtained by placing π_1 *above* π_1, i.e. by decreeing that every element of
π_1 is (strictly) less than every element of π_2, whilst order relations *within*
π_1 and π_2 are as before. [Here *no* identification of least and greatest elements takes place, even if our posets possess such elements.]

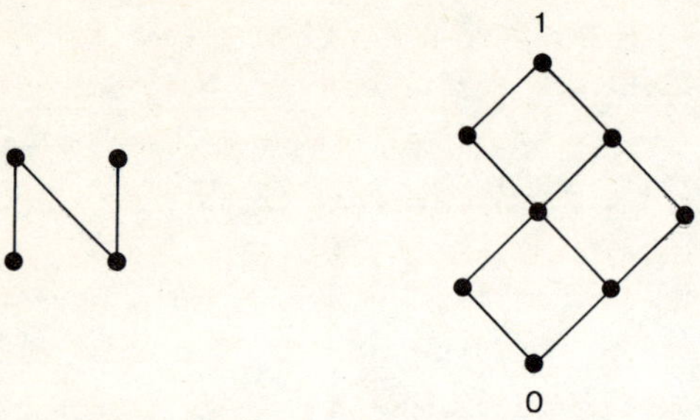

Figure 6

With this convention we can readily discover that $L(\pi_1/\pi_2) = L(\pi_1)/L(\pi_2)$,
where the second slash / now denotes the *lattice ordinal sum* defined in §2
(*with* identification of 0_2 and 1_1). Similarly, if $\pi_1 + \pi_2$ denotes the *car-*
dinal sum [4] of these posets, this being obtained by placing their diagrams
side by side, then $L(\pi_1 + \pi_2) = L(\pi_1) \times L(\pi_2)$. Sets $\{X_t\}$ indexed by the
component posets could now be introduced and the procedure leading to Proposition 3 carried out. The result would be a nesting (resp. crossing) of
the component structures and the details would be as in §2. One consequence
of these remarks is the observation that many simple structures *cannot* be
built up by nesting and crossing from atoms, where an *atom* is a structure
having $A_1 = I$ (identity matrix) and $A_0 = J - I$, singleton poset and 2-element
lattice. The simplest structure which is not built up from atoms has poset
and lattice as in Figure 6.

   We close this paper with an outline of a Type I representation for an
arbitrary, not necessarily finite, distributive lattice L with greatest element 1 and least element 0. We write $\pi = \pi(L)$ for the set of all *prime ideals*
of L and let $\pi_a = \{p \in \pi: a \notin p\}$, $a \in L$. It is well known that $a \rightarrow \pi_a$ is a
lattice homomorphism from L onto a family of subsets of π, and $\pi_0 = \phi$, $\pi_1 = \pi$.
Now let X be the *class of all subsets of* π, and for $a \in L$ define the relation
θ_a on X thus: for $y, z \in X$,

$$(y,z) \in \theta_a \text{ iff } y\Delta z \subseteq \pi'_a$$

where Δ and $'$ denote symmetric difference and complement respectively. It is easy to check that θ_a is an equivalence relation on X and we will see that $\theta_a \,\&\, \theta_b = \theta_{a\vee b}$, $\theta_a \circ \theta_b = \theta_{a\wedge b} = \theta_b \circ \theta_a$, θ_0 = the trivial relation, and θ_1 = the identity relation. The last two assertions are immediate, whilst we see that if $(y,z) \in \theta_a \,\&\, \theta_b$, then $y\Delta z \subseteq \pi'_a \cap \pi'_b = \pi'_{a\vee b}$. Finally, let us suppose that $(x,y) \in \theta_a$ and $(y,z) \in \theta_b$. Then $x\Delta y \subseteq \pi'_a$ and $y\Delta z \subseteq \pi'_b$ whence $x\Delta z = (x\Delta y)\Delta(y\Delta z) \subseteq \pi'_a \cup \pi'_b = \pi'_{a\wedge b}$. On the other hand, if $x\Delta z \subseteq \pi'_{a\wedge b}$ then we take $y = (x \cap \pi_a) \cup (z \cap \pi_b)$ and find that $(x,z) \in \theta_a \circ \theta_b$. Thus $\theta_a \circ \theta_b = \theta_{a\wedge b} = \theta_b \circ \theta_a$ and our assertions are all proved.

4. ACKNOWLEDGEMENTS

The authors are indebted to P. J. Cameron, H. D. Patterson, C. E. Praeger and D. E. Taylor for active participation in joint work, of which the results reported in this paper are a byproduct.

REFERENCES

1. M. Aigner, Combinatorial Theory, Springer-Verlag, New York, 1979.

2. R. A. Bailey, Latin squares with highly transitive automorphism groups. Manuscript, 1978.

3. R. A. Bailey, C. E. Praeger, C. A. Rowley and T. P. Speed, Orbits and permutation characters of generalized wreath products. Manuscript, 1980.

4. G. Birkhoff, Lattice Theory, 3rd ed., American Mathematical Society, Providence, Rhode Island, 1967.

5. R. P. Dilworth and P. Crawley, Algebraic theory of lattices, Prentice-Hall, Inc., Englewood Cliffs, N. J., 1973.

6. G. Grätzer, Lattice Theory. First Concepts and Distributive Lattices, W. H. Freeman and Company, San Francisco, 1971.

7. G. Grätzer, General Lattice Theory, Academic Press, New York, 1978.

8. P. R. Halmos, Reflexive lattices of subspaces, J. London Math. Soc., 4 (Ser. 2) 1971, 257-263.

9. D. G. Higman, Coherent configurations I, Geometriae Dedicata, 4 (1975) 1-32.

10. D. G. Higman, Coherent configurations II, Geometriae Dedicata, 5 (1976) 413-424.

11. D. G. Higman, Lectures on Permutation Representations, Notes taken by Wolfgang Hauptmann, Mathematischen Institut Giessen, 1977.

12. B. Jónsson, On the representation of lattices, Math. Scand., 1 (1953) 193-206.

13. J. A. Nelder, The analysis of randomised experiments with orthogonal block structure, I. Block structure and the null analysis of variance, Proc. Roy. Soc. (London) Ser. A., 273 (1965) 147-162.

14. J. A. Nelder, The analysis of randomised experiments with orthogonal block structure, II. Treatment structure and the general analysis of variance, Proc. Roy. Soc. (London) Ser. A., 273 (1965) 163-178.

15. J. Ogawa, Statistical theory of the analysis of experimental designs, Marcel Dekker, Inc., New York, 1974.

16. D. Raghavarao, Construction and combinatorial problems in design of experiments, John Wiley & Sons, Inc., New York, 1971.

17. S. R. Searle and H. V. Henderson, Dispersion matrices for variance component models, J. Amer. Statist. Assoc., 74 (1979) 465-470.

18. T. N. Throckmorton, Structures of classification data, Ph.D. Thesis, Iowa State University, (1961) iii + 177 pp.

MODULES OVER VALUATION RINGS

L. FUCHS

Department of Mathematics
Tulane University
New Orleans, Louisiana
U.S.A.

This is a brief survey of some fundamental results on modules over valuation rings. For discrete valuation domains, the modules are known to have a strong resemblance to abelian p-groups or modules over the integers localized at a prime, but the non-discrete case turns out to be totally different. This case is being studied by L. Salce and the author, and some of the new results are also incorporated into this survey.

1. PRELIMINARIES

By a *valuation ring* we mean a commutative ring R with 1 whose ideals are totally ordered under inclusion. Equivalently, for each pair a,b of elements in R, either a divides b or vice-versa. The elements of R which are non-units form an ideal P, the unique maximal ideal of R. Thus R is a local ring. If in addition R is a domain, it has a field of quotients, Q. The R-submodules of Q are likewise totally ordered by inclusion; this is ex-pressed by saying that Q is *uniserial*. The R-modules $K = Q/R$ and $K_0 = Q/P$ are uniserial as well.

Let $V = R\backslash P$ be the group of units of R. The factor group $\Gamma = Q^\times/V$ of the multiplicative group $Q^\times$ of Q mod V is the *group of divisibility* of R. It can be totally ordered by setting

$$a + V \leq b + V \text{ if and only if } Rb \leq Ra$$

In this way, Γ becomes a totally ordered group, and the function $v: Q^\times \to \Gamma$ (acting as $v(a) = a + V$) is a valuation of Q. Conversely, if Q is a field with a non-archimedean (exponential) valuation v in a totally ordered abelian group Γ, then $R = \{a \in Q: v(a) \geq 0\}$ is a valuation domain with $P = \{a \in Q: v(a) > 0\}$ as maximal ideal, and the divisibility group of R is the image of v in Γ. Note that every totally ordered abelian group Γ is the divisibility group of a suitable valuation domain.

A valuation ring R is said to be *maximal* if an infinite system of congruences

$$x \equiv r_i \pmod{L_i}$$

where $i \in I$, $r_i \in R$, L_i are ideals of R and I is any index set, has a solution in R whenever each of its finite subsystems is solvable in R. This is the same thing as saying that R is *linearly compact* in the discrete topology. (It is then linearly compact in the R-topology where the non-zero principal ideals form a system of open neighbourhoods of 0.)

If R/L is linearly compact for all non-zero ideals L of R, then R is called *almost maximal*. For instance, the ring $\mathbb{Z}_p$ of integers localised at p is an almost maximal valuation domain, while its completion $\mathbb{Z}_p^*$, the ring of the p-adic integers, is a maximal valuation domain. Gill [3] points out that an almost maximal valuation ring that is not a domain has to be maximal. (It is still an open question whether every valuation ring is a quotient of a valuation domain.)

All rings considered here are supposed to have an identity and are commutative. Unless stated otherwise, R *will always denote a valuation ring*, and only unital R-modules will be considered. The symbol $\leq$ between modules will mean inclusion, while < will be reserved for proper inclusion. If M is an R-module and $a \in M$, then the *annihilators*

$$\text{Ann } M = \{r \in R: rM = 0\}, \quad \text{Ann } a = \{r \in R: ra = 0\}$$

are ideals of R. If R is a domain, M is called *torsion* if Ann $a \neq 0$ for every $a \in M$, and *torsion-free* if Ann $a = 0$ for every $a \in M$, $a \neq 0$.

Let L be an ideal of R and M an R-module. We define LM as the submodule of M generated by all rx ($r \in L$, $x \in M$); R being a valuation ring, this set is already a submodule.

With every R-module M we associate two vector spaces over the residue field R/P, the *socle* and the *top* of M defined as follows:

Soc M = {$a \in M$: Ann $a \geq P$}

and

Top M = M/PM

Notice that Soc (M/Soc M) = 0 and Top PM = 0 whenever P is not a principal ideal (R is a *non-discrete* valuation ring). A major difference between abelian groups and modules over non-discrete valuation rings R lies exactly in the fact that a non-zero torsion R-module may have trivial socle and a non-divisible R-module may have trivial top.

2. PROJECTIVE MODULES

We start the study of R-modules with the projective R-modules. We have already noticed above that valuation rings are special cases of local rings. For local rings, Kaplansky [5] has shown that all projective modules are free. Hence:

THEOREM 1. Projective modules over valuation rings are free.

It is easy to see that in any direct decomposition of a free R-module F into copies of R, the number of summands is equal to the dimension of Top F as an R/P-vector space, so this number is uniquely determined by F.

The question of global projective dimensions of valuation rings presents itself. If a ring is not Artinian semisimple, then its left global dimension is known to be equal to the supremum of the projective dimensions of its left ideals plus 1. Osofsky [9] has shown that the projective dimension of an ideal L in a valuation domain R can be computed as follows. Let ω_n denote the first ordinal of cardinality $\aleph_n$; we let $n \geq -1$ and set $\omega_{-1} = 0$. We can write L = $\cup${Ra: $a \in L$}. Suppose we can select a well-ordered chain of principal ideals Ra of order type cofinal with ω_n whose union is L. If n is the smallest such integer, then the projective dimension of L equals n + 1.

Hence the global dimension of R will be $n + 2$, if n is the largest integer occurring for ideals of R. For instance, if R is a non-discrete valuation domain in which every ideal is countably generated, then the global dimension of R equals 2.

It is worth-while pointing out that valuation domains are Prüfer domains, and therefore an R-module is flat if and only if it is torsion-free.

3. DIVISIBLE AND INJECTIVE MODULES

An R-module M is called *divisible* if $rM = M$ for every non-zero $r \in R$. Evidently, Q and all its quotients are divisible.

In every R-module M, there exists a maximum divisible submodule dM such that M/dM has no non-zero divisible submodule. In general, dM is not a summand of M. (Take, for instance, a cyclic extension of a direct sum of countably many copies of an indecomposable injective in its injective hull.)

For torsion-free modules over integral domains, divisibility and injectivity are equivalent. Hence torsion-free injectives are exactly the vector spaces over Q, and therefore direct sums of copies of Q.

For a valuation ring R, there is only one simple R-module, namely, R/P. Its injective hull $E(R/P)$ is (the smallest) cogenerator of the category of R-modules. In the domain case, $E(R/P)$ must contain a copy of $Q/P = K_0$, since R/P is an essential submodule of K_0. Thus $E(R/P) = E(K_0)$.

In general, K_0 is a proper submodule of its injective hull. The case when K_0 is injective has been characterised by Matlis [7]:

THEOREM 2. For a valuation domain R, K_0 is injective if and only if R is almost maximal.

This result has been generalised by Gill [3].

THEOREM 3. For a local ring R, the following are equivalent:

(a) $E(R/P)$ is uniserial

(b) uniform R-modules are uniserial

(c) R is an almost maximal valuation ring

Here uniformity means that the module does not contain two non-zero submodules with 0 intersection. Note that, in the case of domains, (a) implies that Q/P is injective and (b) implies that all Q/L (with L an ideal of R) are injective.

For arbitrary valuation domains R, the indecomposable injectives are of the form E(R/L) with L an ideal of R. Matlis [7] has shown that

$$E(R/L) \simeq E(R/J) \text{ if and only if } L \simeq J$$

where L,J are ideals of R. Notice that this happens exactly if L = Ja for some a ≠ 0 in Q.

Thus, if $\Gamma = \mathbb{R}$, then the indecomposable injectives are: Q, E(Q/R) and E(Q/P) (these are pairwise non-isomorphic). However, if $\Gamma = \mathbb{Q}$, then there are continuously many pairwise non-isomorphic ideals of R, and consequently, there are continuously many non-isomorphic indecomposable injectives.

As the injective hull of a cyclic module over a valuation ring R is indecomposable, it is clear that every element of an injective R-module M is contained in an indecomposable summand of M. Hence we get at once the first part of the following theorem, due to Warfield [15]:

THEOREM 4. Every injective module over a valuation ring is the injective hull of a direct sum of indecomposable injectives. This direct sum is unique up to isomorphism.

The second statement is a consequence of general theorems of Warfield, using the fact that the endomorphism rings of indecomposable injectives are local rings. Theorem 4 yields a satisfactory description of injectives by means of cardinal invariants.

A straightforward application of Baer's criterion of injectivity shows that the torsion submodule of an injective R-module is injective, so it splits off as a summand. This reduces the study of injectives to the torsion case. It is worth-while noticing that if T is a torsion, injective R-module, then we can write $T = T_0 \oplus T_1$ where Soc T_0 = Soc T is essential in T_0, and Soc T_1 = 0. (Just choose T_0 to be an injective hull of Soc T.)

Turning to *quasi-injective* R-modules (defined as fully invariant submodules of injectives), the following result should be pointed out (see Vámos [11]):

THEOREM 5. A valuation ring R is almost maximal if and only if all submodules
of E(R/P) are quasi-injective.

Notice that if L is an arbitrary non-zero ideal of R, a submodule of Q/L
need not be quasi-injective, even if R is almost maximal. An easy criterion
can be established for J/L (with R-submodules $J, L \neq 0$ of Q) to be quasi-
injective.

THEOREM 6. Let R be an almost maximal valuation domain. For proper submodules $0 \neq L < J$ of Q, J/L is quasi-injective if and only if $L : L \leq J : J$.

Here $L : J = \{x \in Q: xJ \leq L\} \cong \text{Hom}(J,L)$. In view of Lemma 1 (see section 5
below), Theorem 6 covers all uniserial torsion modules which are not injective. (The non-torsion case is trivial: all quasi-injectives are injective.)

It is clear that for every injective R-module M, the submodules

$$M[L] = \{x \in M: \text{Ann } x \geq L\}$$

and

$$M^+[L] = \{x \in M: \text{Ann } x > L\}$$

for any ideal L of R, are quasi-injective.

4. FINITELY GENERATED AND FINITE RANK MODULES

We begin with a special case: the *finitely presented* modules M. That is,
$M \cong F/N$ with F finitely generated free and N finitely generated. If in addition both F and N can be chosen to have a single generator then M is *cyclically presented*.

The following result by Warfield [16] totally describes what finitely
presented R-modules are like.

THEOREM 7. A module over a valuation ring is finitely presented if and only
if it is a finite direct sum of cyclically presented modules.

Turning our attention to finitely generated modules over valuation
domains R, Kaplansky [4] proved that finitely generated torsion-free modules
are free. Consequently, the torsion submodule of a finitely generated

R-module splits off as a summand. For finitely generated torsion modules he
proved that they are direct sums of cyclic modules whenever R is almost maximal. Matlis [8] (for valuation domains R), Lafon [6] and Gill [3] have
shown independently:

THEOREM 8. Let R be a local ring. All finitely generated R-modules are
direct sums of cyclic R-modules exactly if R is an almost maximal valuation
ring.

For a general discussion of commutative rings over which finitely generated modules are direct sums of cyclics, we refer to Brandal [1].

Matlis [7] points out that, over every valuation domain, finitely generated submodules of a direct sum of quotients of Q are again direct sums of
cyclics.

We say that a module M is *of finite rank* if M can be embedded in a
finite direct sum of uniform modules, or equivalently, E(M) is a finite direct
sum of indecomposable injectives. These modules can be characterised in a
satisfactory way provided that the ring is almost maximal (cf. also Fleischer
[2]).

THEOREM 9. If R is an almost maximal valuation domain, then every torsion
R-module of finite rank is a direct sum of a finite number of uniserial R-
modules.

From Lemma 3 (see section 6 below) it will follow that all these uniserial modules are pure-injective. Using this, it follows that for torsion
modules of countable rank we have:

THEOREM 10. Let R be an almost maximal valuation domain. A torsion R-module
of countable rank is a direct sum of uniserial R-modules if and only if it is
the union of an ascending sequence of pure submodules of finite rank.

5. UNISERIAL MODULES AND THEIR DIRECT SUMS

It is evident that both submodules and factor modules of uniserial modules
are uniserial again. Hence quotients of Q and their submodules are uniserial
modules. It is straightforward to check that if R is an almost maximal valuation domain, then these are the only uniserial R-modules.

LEMMA 1. Uniserial modules over an almost maximal valuation domain R are of the form J/L where $0 \leq L < J \leq Q$ are R-submodules of Q. Moreover, $J/L \simeq J'/L'$ if and only if, for some non-zero $a \in Q$, $L' = aL$ and $J' = aJ$.

Following Salce [10], for a non-zero ideal L of R we define

$$L* = \{a \in Q: aL < L\}$$

This is always a prime ideal of R that contains all proper ideals of R isomorphic to L. The ring End L of all R-endormorphisms of a non-zero ideal L of R turns out to be the ring R_{L*} (that is, R localised at the prime ideal L*).

In the study of direct sums of uniserial modules, the following Lemma is crucial.

LEMMA 2. Let R be an almost maximal valuation domain and $U \simeq J/L$ a uniserial R-module with $L \neq 0$. Then, End U is the completion of S = End L in its S-topology if J = Q, and is isomorphic to (End J)/(L : J) if $J \neq Q$. In both cases, End U is a valuation ring.

Here L : J is an ideal of End J $\simeq R_{J*}$. A consequence of the last result is that uniserial modules over an almost maximal valuation domain have the Exchange Property. In addition, uniserial R-modules U are *countably small* in the sense that if $\psi: U \to \oplus M_i$ with arbitrary modules M_i and i ranging over an arbitrary index set I, then ψ factors through $\oplus\{M_i : i \in I_0\}$ with a suitable countable subset I_0. Hence, in view of Lemma 2, results by Warfield [15] imply:

THEOREM 11. Let R be an almost maximal valuation ring and $M = \oplus\{U_i : i \in I\}$ a direct sum of uniserial R-modules. Then, this decomposition of M is unique up to isomorphism, and every summand of M is isomorphic to a direct sum $\oplus\{U_i : i \in J\}$ for some subset J of I.

The following example shows that submodules of direct sums of uniserial modules need not be again direct sums of uniserial modules (not even if purity is assumed).

Let R be an almost maximal valuation domain with value group $\mathbb{R}$, and let T be the R-module generated by $a_0, a_1, \ldots, a_n, \ldots$ subject to the relations $r_0 a_0 = 0$ and $r_n a_n = a_0$ for $n = 1, 2, \ldots$ where the sequence $v(r_n)$, $r_n \in R$,

tends increasingly to a real number $\alpha > 0$. This T is obviously not a direct
sum of uniserial modules (look at a_0). Let U be a uniform R-module isomorphic to P/Rs where $s \in R$ has value $v(r_0) + \alpha$. The direct sum M of T and
U with amalgamated submodules Ra_0 and Rsr_0^{-1}/Rs will be isomorphic to
$U \oplus \oplus\{R/Rr_n : n \geq 1\}$ with T pure in M.

6. PURITY

Recall that a submodule N of M is called *pure* if the canonical map
$N \otimes_R A \to M \otimes_R A$ is injective for every R-module A. Warfield [14] has proved
that for valuation rings R, N is pure in M if and only if $rN = N \cap rM$ for all
$r \in R$. It is readily seen that a submodule N of a torsion-free module M over
a valuation domain is pure exactly if M/N is torsion-free.

   This notion of purity has properties analogous to purity over principal
ideal rings. Of the numerous aspects of the corresponding relative homological algebra, we discuss only the *pure-projectives* and the *pure-injectives*
(these are defined to have the projective and the injective property, respectively, relative to all pure-exact sequences).

   For any ring R whatsoever, the pure-projectives are precisely the summands of direct sums of finitely presented R-modules. These direct sums are
by Theorem 7, just direct sums of cyclically presented modules and the same
is true for their summands (Warfield [14]):

THEOREM 12. The pure-projective modules over a valuation ring are exactly
the direct sums of cyclically presented modules.

   The characterisation of pure-injectives is more difficult. Satisfactory
results are known only in the torsion-free case. Warfield [14] has shown:

THEOREM 13. Every pure-injective module over a valuation domain R is the
direct sum of an injective R-module and a pure-injective R-module without
elements of infinite height.

THEOREM 14. A torsion-free pure-injective R-module without elements of
infinite height is the pure-injective hull of a direct sum of R-modules
isomorphic to ideals of R. This direct sum is unique up to isomorphism.

Recall that an R-module M has no elements of infinite height means $\cap\{rM: r \in R, r \neq 0\} = 0$.

There are situations where R itself (as an R-module) is pure-injective; it is well known that this is the case for complete discrete valuation domains. As is shown by Warfield [14], this is true if and only if R is linearly compact in the discrete topology. Moreover, he proved:

THEOREM 15. Let R be a valuation domain and S a maximal immediate extension of R. The pure-injective hull E of a finitely generated R-module M satisfies $E \cong M \otimes_R S$. It is a direct sum of a finite number of cyclic S-modules.

More can be said of pure-injectivity if R happens to be almost maximal.

LEMMA 3. Every uniserial torsion module over an almost maximal valuation ring is pure-injective.

By making use of this Lemma, the usual proof for abelian groups carries over, mutatis mutandis, to establish (cf. Salce [10]):

THEOREM 16. Let R be an almost maximal valuation ring. An R-module is pure-injective if and only if it is a summand of a direct product of uni-serial torsion R-modules (all these can be assumed to have non-trivial socles).

Fleischer [2] introduced a different kind of purity which we shall call "copure" (this is in accordance with the terminology of Walker [12]). A submodule N of M is said to be *copure* in M if every coset $\bar{a} \in M/N$ can be represented by an element $a \in M$ such that Ann a = Ann $\bar{a}$. Equivalently, for every cyclic R-module C, the canonical map $\text{Hom}_R(C,M) \to \text{Hom}(C,M/N)$ is surjective. Since purity can be characterised in the same way by using cyclically presented R-modules C, it is evident that copurity is stronger than purity.

As is pointed out by Salce [10], the analogue of Theorem 12 holds true:

THEOREM 17. The copure-projective modules over a valuation ring R are exactly the direct sums of cyclic R-modules. Such decompositions are unique up to isomorphism.

The class of copure-injectives includes the class of pure-injective modules; it is not known whether or not these classes coincide.

Just as in abelian group theory, purity serves as a main tool in establishing direct summands. For valuation rings, however, one is confronted with various kinds of difficulty. First, we have been unable to establish the existence of pure submodules of finite rank in torsion modules. Secondly, it can very well happen that a proper pure submodule of a bounded torsion module is essential.

In conclusion, we mention a lemma is useful in finding pure submodules.

LEMMA 4. For any R-module M, a $\in$ M is an element with maximal annihilator in the coset a + PM if and only if Ra is pure in M.

REFERENCES

1. W. Brandal, Commutative rings whose finitely generated modules decompose, Lecture Notes in Math. 723, Springer, 1979.

2. I. Fleischer, Modules of finite rank over Prüfer rings, Ann. of Math. 65 (1957), 250-254.

3. D. T. Gill, Almost maximal valuation rings, J. London Math. Soc. 4 (1971), 140-146.

4. I. Kaplansky, Modules over Dedekind rings and valuation rings, Trans. Amer. Math. Soc. 72 (1952), 327-340.

5. I. Kaplansky, Projective modules, Ann. of Math. 68 (1958), 372-377.

6. J. P. Lafon, Sur un problème d'Irving Kaplansky, C. R. Acad. Sci. Paris, Sér. A, 268 (1969), 1309-1311.

7. E. Matlis, Injective modules over Prüfer rings, Nagoya Math. J. 15 (1959), 57-69.

8. E. Matlis, Decomposable modules, Trans. Amer. Math. Soc. 125 (1966), 147-179.

9. B. L. Osofsky, Global dimension of valuation rings, Trans. Amer. Math. Soc. 127 (1967), 136-149.

10. L. Salce, Moduli di torsione su domini di valutazione, Notes, Conference on Abelian Groups and Modules, Trento, May 1980.

11. P. Vámos, Classical rings, J. Algebra 34 (1975), 114-129.

12. C. P. Walker, Relative homological algebra and abelian groups, Ill. J. Math. 10 (1966), 186-209.

13. R. B. Warfield, Jr., A Krull-Schmidt theorem for infinite sums of modules, Proc. Amer. Math. Soc. 22 (1969), 460-465.

14. R. B. Warfield, Jr., Purity and algebraic compactness for modules, Pac. J. Math. 28 (1969), 699-719.

15. R. B. Warfield, Jr., Decompositions of injective modules, Pac. J. Math. 31 (1969), 263-276.

16. R. B. Warfield, Jr., Decomposability of finitely presented modules, Proc. Amer. Math. Soc. 25 (1970), 167-172.

WHEN ARE RADICAL CLASSES OF ABELIAN GROUPS
CLOSED UNDER DIRECT PRODUCTS?

B. J. GARDNER

Department of Mathematics
University of Tasmania
Hobart, Tasmania
Australia

A *radical class* of abelian groups is a non-empty class R satisfying the
following conditions:

(1) R is homomorphically closed

(2)   For every $A \in R$, $R(A) \in R$ where $R(A) = \Sigma\{S \leq A: S \in R\}$

(3) $R(A/R(A)) = 0$ for every $A \in R$

Equivalently, a radical class is a non-empty class of abelian groups which
is closed under homomorphic images, direct sums and extensions.

Radical classes have elsewhere been called *torsion classes*, but it
seems more appropriate to reserve this term for radical classes which are
hereditary (closed under subgroups). This is also the convention adopted
in the book by Mishina and Skornyakov [12], a good source of information on
radical classes of modules in general, in which the above, and other,
characterizations of radical classes can be found.

Every non-empty class M of abelian groups is contained in a smallest
radical class $L(M)$, the class of abelian groups A such that for every non-
zero homomorphic image B of A, we have $\mathrm{Hom}(M,B) \neq 0$ for some $M \in M$ [4].

Here are some examples of radical classes of abelian groups.

(a) T , the class of all torsion groups

(b) T_p, the class of all p-groups, where p is a prime

(c) T_S, the class of all direct sums of p-groups for p in a fixed set S of primes

(d) $\mathcal{D}$, the class of all divisible groups

(e) $\mathcal{D}_p$, the class of all p-divisible groups, where p is a prime

(f) $\mathcal{D}_S$, the class of all groups divisible by each p in a fixed set S of primes

1. DIRECT PRODUCTS

It will be readily observed that the classes in (d),(e) and (f) in the list above are closed under arbitrary direct products. Indeed they are the only radical classes of abelian groups known to be so closed. This brings us to the main subject of this paper. (We note in passing that not too much is known about direct product closure in radical theory in general. Radical classes of rings, (non-abelian) groups and various other structures are defined and characterized by conditions similar to, but in general less neat than, those used above for radical classes of abelian groups. A good reference is [16].)

Previously we obtained the following intriguing but not completely satisfying result (see [5], Theorem 7.2).

THEOREM 1.1. A radical class R of abelian groups is closed under countable direct products if and only if $R = L(M)$ for a class M of torsion-free groups.

We shall not delve into the proof of this theorem here; suffice it to say that the countability involved is a consequence of Hulanicki's famous result [10] that for any sequence $A_1, A_2, A_3, \ldots$ of abelian groups, $\Pi A_n / \oplus A_n$ is algebraically compact.

In [5] we stated the question: are there any radical classes other than the $\mathcal{D}_S$ which are closed under arbitrary direct products? Recently Ivanov [11] has taken up the matter, showing, in effect, that for every radical class R, the class $\hat{R}$ of torsion-free groups in R is closed under countable direct products. No further examples were given, however, of radical classes R for which $\hat{R}$ is closed under *arbitrary* direct products.

In fact, there are plenty of radical classes closed under *uncountable* direct products. For the terminology in the next result we refer to Fuchs [3], Vol. II.

THEOREM 1.2. Let S be a class of slender groups, and let

$$U(S) = \{A: \text{Hom}(A,S) = 0 \text{ for all } S \in S\}$$

Then $U(S)$ is a radical class closed under direct products of the form $\Pi\{A_i: i \in I\}$, where $|I|$ is non-measurable.

Proof.   The exactness properties and behaviour towards direct sums of $\text{Hom}(\_,\_)$ make $U(S)$ a radical class (cf. [12]). If $\{A_i: i \in I\} \subseteq U(S)$ and $|I|$ is non-measurable, then any homomorphism $f: \Pi A_i \to S \in S$ annihilates $\oplus A_i$, so by a Theorem of Łoś ([3], Vol. II, Theorem 94.4) for torsion-free A_i, generalized to arbitrary groups A_i by Göbel ([7], Lemma 3.6), we have $f = 0$, whence $\Pi A_i \in U(S)$.

Every countable reduced torsion-free abelian group is slender, (Sasiada [15]) so Theorem 1.2 provides lots of examples of radical classes. The status of measurable cardinals is uncertain.

In view of the central role thus far played by the classes $\mathcal{D}_S$, a possible approach to our problem is to find diverse characterizations of $\mathcal{D}_S$ and then examine the extent to which product-closed radical classes in general mimic the properties involved in these characterizations. In the sequel we shall be interested in the following descriptions of $\mathcal{D}_S$.

(1) $\mathcal{D}_S$ is the class of groups A for which there is a short exact sequence $0 \to X \to A \to Y \to 0$ where $X \in \mathcal{D} \cap T_S = L(\{Z(p^\infty): p \in S\}$ and Y is a module over the ring $Q(S) = \{m/n: m,n \in Z \text{ and if } p|n \text{ then } p \in S\}$

(2) $\mathcal{D}_S$ is the class of relative injectives for the short exact sequences $0 \to X' \to X \to X'' \to 0$ where X'' is in T_S (see Hilton and Yahya [9], 4.2 Theorem)

2. BOUNDEDNESS

Let R be a radical class of abelian groups, α a cardinal number. We shall say that R is *bounded* by α if for every $a \in A \in R$, there exists a subgroup B of A such that $a \in B \in R$ and $|B| \le \alpha$. Some examples:

(a) T, or more generally, T_S, is bounded by $\aleph_0$ since for every $a \in A \in T_S$ we have $a \in \langle a \rangle \in T_S$

(b) $\mathcal{D}$ is bounded by $\aleph_0$ since for $a \in A \in \mathcal{D}$ we have $a \in E(\langle a \rangle) \in \mathcal{D}$, where $E(\ )$ indicates an injective hull

(c) Every radical class closed under pure subgroups - see [4] for a characterization of such classes - is bounded by $\aleph_0$ since every group element is contained in a countable pure subgroup (see [3], Vol. I, p. 115)

(d) For a ring R with identity, let P be a projective R-module and R_P the class of R-modules M generated by submodules which are homomorphic images of finite direct sums of copies of P. Then R_P is bounded by $|P|\aleph_0$

It is not known whether there are any unbounded radical classes of abelian groups (or modules in general). If there are no such classes then bounding cardinals have to be very large. If G is a torsion-free *rigid* group (i.e. Hom(G,G) has rank 1) then L(G) is not bounded by any cardinal $\alpha < |G|$ since if $a \in A \leq G$ and $|A| < |G|$ then Hom(G,A) must be zero, whence $A \not\in L(G)$. There exist rigid groups of very large cardinalities ([3] Vol. II, p. 130).

If R is a radical class of *rings* we say that R is bounded by α if for every $a \in A \in R$ there exists an *ideal* B of A such that $a \in B \in R$ and $|B| \leq \alpha$. An example of an *unbounded* radical class of rings is the smallest radical class containing all simple rings, since simple rings can be arbitrarily large, and if S is simple with $|S| > \alpha$, no $a \in S - \{0\}$ is contained in an ideal of cardinality less than α.

We are interested in the question: does closure under countable direct products imply closure under arbitrary direct products? Analogous questions can also be asked for uncountable direct products. The next result provides an answer to certain such questions for bounded radical classes. The Theorem is derived from a result of Nunke ([13], Theorem 2.2).

THEOREM 2.1. Let $R \neq \{0\}$ be a radical class of abelian groups, α an infinite cardinal number. If R is bounded by α and closed under direct products of the form $\Pi\{A_i : i \in I\}$ where $|I| \leq 2^{(2^\alpha)}$, then there is a short exact sequence

$$0 \to Z \overset{f}{\to} G \to H \to 0$$

where G is in R, such that for every abelian group B we have $R(B) = \text{Im}(g)$ in the induced diagram

$$\text{Hom}(G,B) \longrightarrow \text{Hom}(Z,B) \longrightarrow \text{Ext}(H,B)$$

$$g\downarrow \qquad \cong\downarrow \qquad \uparrow$$

$$B \;\underset{=}{\rightarrow}\; B \;\underset{=}{\rightarrow}\; B$$

In this event R is closed under arbitrary direct products.

Proof. Let C be a set of groups consisting of one isomorphic copy of each group in R of cardinality at most α, and let $C = \Pi\{X: X \in C\}$.

Since R is bounded by α, if $a \in A \in R$ then for some $X \in C$ there is an injective homomorphism $\xi_X: X \to A$ with $a \in \text{Im}(\xi_X)$.

For each cardinal β let $\Gamma(\beta)$ denote the number of groups of order β, so $\Gamma(\beta) \leq \aleph_0$ if β is finite and $\Gamma(\beta) \leq 2^\beta$ otherwise. However $\Gamma(\beta) \leq \Gamma(\alpha)$ for every $\beta \leq \alpha$, so we have

$$\Sigma\{\Gamma(\beta): \beta \leq \alpha\} \leq \alpha\Gamma(\alpha) \leq \alpha 2^\alpha = 2^\alpha$$

Hence $|C| \leq 2^\alpha$, so $C \in R$. Now define $\xi: C \to A$ to be $\pi_X\xi_X$ where π_X is the natural projection from C onto X. We have $a \in \text{Im}(\xi_X) = \text{Im}(\xi)$. Thus for every $a \in A \in R$ there is a $\xi \in \text{Hom}(C,A)$ such that $a \in \text{Im}(\xi)$.

Let $G = \Pi\{C_v: v \in C\}$ where each C_v is C. Now

$$|C| = \Pi\{|X|: X \in C\} \leq \alpha^{|C|} = \alpha^{(2^\alpha)}$$

and

$$2^{(2^\alpha)} \leq \alpha^{(2^\alpha)} \leq (2^\alpha)^{(2^\alpha)} = 2^{(2^\alpha)}$$

since $2 < \alpha < 2^\alpha$. Thus $|C| \leq 2^{(2^\alpha)}$ and so $G \in R$.

Let $x = (c_v) \in G$ be defined by $c_v = v$ for all $v \in C$. With the notation previously used, $a = \xi(v_0)$ for some $v_0 \in C$. Denoting by $\pi_v: G \to C_v$ the co-ordinate projection, we get

$$a = \xi(v_0) = \xi(c_{v_0}) = \xi\pi_{v_0}((c_v)) = \xi\pi_{v_0}(x)$$

It follows that

$$A = \{\eta(x): \eta \in \text{Hom}(G,A)\} \tag{*}$$

We define $f: Z \to G$ by $f(1) = x$. Since R is closed under *some* infinite direct products, it contains a non-zero torsion-free group K. If y is a non-zero element of K, then by (*) there is an $\eta \in \text{Hom}(G,K)$ such that $\eta(x) = y$. Suppose $f(m) = 0$ for some $m \in Z$. Then $my = m\eta(x) = m\eta f(1) = \eta f(m) = 0$, so $m = 0$. Thus f is injective.

Now consider the exact sequence

$$0 \to Z \xrightarrow{f} G \to H \to 0$$

where $H = G/f(Z)$. For any abelian group B, we have an induced exact sequence

$$0 \to \mathrm{Hom}(H,B) \to \mathrm{Hom}(G,B) \xrightarrow{f^*} \mathrm{Hom}(Z,B) \to \mathrm{Ext}(H,B) \to \mathrm{Ext}(G,B) \to 0$$

and maps $g: \mathrm{Hom}(G,B) \to B$ and $h: \mathrm{Hom}(Z,B) \cong B$ such that $g = f^*h$. Since $R(B) \in R$, we have

$$R(B) = \{\eta(x): \eta \in \mathrm{Hom}(G,R(B))\} = \{\eta(x): \eta \in \mathrm{Hom}(G,B)\}$$

(since G is in R). But for $\eta \in \mathrm{Hom}(G,B)$, we have

$$g(\eta) = hf^*(\eta) = h(\eta f) = \eta f(1) = \eta(x)$$

and so

$$R(B) = \{\eta(x): \eta \in \mathrm{Hom}(G,B)\} = \mathrm{Im}(g)$$

If now $\{A_i: i \in I\} \subseteq R$, where I is a set, then we have a commutative, exact diagram

$$\begin{array}{ccccc}
\mathrm{Hom}(G,\Pi A_i) & \xrightarrow{f^*} & \mathrm{Hom}(Z,\Pi A_i) & \xrightarrow{e} & \mathrm{Ext}(H,\Pi A_i) \\
\downarrow{i_1} & & \downarrow{i_2} & & \downarrow{i_3} \\
\Pi\mathrm{Hom}(G,A_i) & \xrightarrow{j} & \Pi\mathrm{Hom}(Z,A_i) & \xrightarrow{k} & \Pi\mathrm{Ext}(H,A_i)
\end{array}$$

where $\mathrm{Im}(f^*) \in R$, $R(\mathrm{Im}(e)) = 0$ and the vertical maps are isomorphisms. Hence $\mathrm{Im}(j) \in R$ and $R(\mathrm{Im}(k)) = 0$, so $\mathrm{Im}(j) = R(\Pi\mathrm{Hom}(Z,A_i))$. It follows that $R(\Pi A_i) = \Pi R(A_i)$.

From now on we shall describe the situation of Theorem 2.1 by saying that R is *represented* by the sequence

$$0 \to Z \xrightarrow{f} G \to H \to 0$$

PROPOSITION 2.2. For any non-empty set S of primes, $\mathcal{D}_S$ is represented by the sequence $0 \to Z \to Q(S) \to Q(S)/Z \to 0$.

Proof. For any group A, $\mathrm{Hom}(Q(S),A)$ is S-divisible, while $\mathrm{Ext}(Q(S)/Z,A) \cong \Pi\{\mathrm{Ext}(Z(p^\infty),A): p \in S\}$ is S-reduced (see, e.g. [8] Proposition 44(a), pp. 35-36). Hence in the exact sequence

$$\mathrm{Hom}(Q(S),A) \to \mathrm{Hom}(Z,A) \to \mathrm{Ext}(Q(S)/Z,A)$$

the first map has an S-divisible image, which must be $\mathcal{D}_S(\mathrm{Hom}(Z,A))$, since the second map has an S-reduced image.

Proposition 2.2 indicates that a radical class can be represented by more than one exact sequence, and that in particular, it is not always necessary to use the whole group G of the proof of Theorem 2.1.

We next look at the problem of determining which sequences

$$0 \to Z \to G \to H \to 0$$

represent radical classes. We first dispose of the case where H is torsion, and for this we need to look at $R(A)$ for a radical class R and a *cotorsion* group A. In what follows, I_p denotes the group or ring of p-adic integers.

PROPOSITION 2.3. Let $A = \Pi A(p)$, where for each prime p, $A(p)$ is a reduced, torsion-free, algebraically compact p-adic group, let R be a radical class closed under countable direct products and let $S = \{p: I_p \notin R\}$. Then

$$R(A) = \mathcal{D}_S(A) = \Pi\{A(p): p \notin S\}$$

Proof.   Since

$$A/\Pi R(A(p)) = \Pi A(p)/\Pi R(A(p)) \simeq \Pi[A(p)/R(A(p))]$$

we have $R(A/\Pi R(A(p))) = 0$ (see, e.g. [12]); therefore, since $\Pi R(A(p)) \in R$, we also have $R(A) = \Pi R(A(p))$. If $I_p \in R$ then $A(p)$, as an I_p-module and thus a homomorphic image of a direct sum of copies of I_p, is in R. If $I_p \notin R$, then $R(I_p)$ is pure in I_p ([6], Proposition 1.1) so either $R(I_p) = 0$ or $I_p/R(I_p)$ is non-zero and divisible ([2], Theorem 1) and hence in R ([6], Corollary 2.3). The latter is impossible, so $R(I_p) = 0$. But then $\mathrm{Hom}(R(A(p)),I_p = 0$, so $R(A(p))$ is p-divisible ([2], p. 52) and hence zero. Thus we have

$$R(A) = \Pi\{A(p): I_p \in R\} = \Pi\{A(p): p \notin S\}$$

while

$$\mathcal{D}_S(A) = \Pi\{A(p): p \notin S\}$$

PROPOSITION 2.4. Let A be a reduced, adjusted, cotorsion group, R a radical class closed under countable direct products, $S = \{p: Z(p) \notin R\}$. Then $R(A) = \mathcal{D}_S(A)$.

Proof.   We have $A = \Pi A(p)$ where each $A(p)$ is a reduced, adjusted, co-torsion, p-adic group (see [8], pp. 50-51). Let T_p be the torsion subgroup of $A(p)$ (necessarily a reduced p-group). If $p \in S$ then $A(p)$ is p-reduced and so $R(A(p)) \subseteq \mathcal{D}_S(A(p)) = 0$. If $p \notin S$ then $T_p \in R$ while, since R contains torsion-free groups, the divisible group $A(p)/T_p$ is also in R ([6], Corollary 2.3), whence $A(p)$ is in R. Thus (as in the previous proof) $R(A) = \Pi R(A(p)) = \Pi\{A(p): p \notin S\} = \mathcal{D}_S(A)$.

We can now state our general theorem on radicals of cotorsion groups.

THEOREM 2.5. Let R be a radical class closed under countable direct products, and let S be the set of primes p for which every group in R is p-divisible. Then $R(M) = \mathcal{D}_S(M)$ for every cotorsion group M.

Proof.   Let M be cotorsion. Then $M = A \oplus B \oplus C$ where A is divisible, B is a product of reduced torsion-free algebraically compact p-adic groups and C is a reduced, adjusted, cotorsion group.

Now $S = \{p: Z(p) \notin R\}$. If $p \in S$ then clearly $I_p \notin R$, while if $p \notin S$ then R contains a p-reduced group and hence a p-reduced torsion-free group X (since $R = L(M)$ for a class M of torsion-free groups). However since $\text{Hom}(X, I_p) \neq 0$ ([2], p. 52) we then have $R(I_p) \neq 0$ and thus, as in the proof of Proposition 2.3, $I_p \in R$. Thus also $S = \{p: I_p \notin R\}$.

We now have, using Propositions 2.3 and 2.4,

$$R(M) = R(A) \oplus R(B) \oplus R(C) = A \oplus \mathcal{D}_S(B) \oplus \mathcal{D}_S(C)$$

$$= \mathcal{D}_S(A) \oplus \mathcal{D}_S(B) \oplus \mathcal{D}_S(C) = \mathcal{D}_S(M)$$

Theorem 2.5 describes the effect of product-closed radicals on a significant class of groups. In the absence of a complete classification of the product-closed radical classes, further such "local solutions" of the problem seem to be worth seeking.

Our results for cotorsion groups allow us to extract some further information.

THEOREM 2.6. If a radical class R of abelian groups is represented by a short exact sequence $0 \rightarrow Z \rightarrow G \rightarrow H \rightarrow 0$ with H torsion then $R = \mathcal{D}_S$ for some set S of primes.

Proof.   The correspondence $A \to R(A)$ defines a cotorsion functor in the sense of Nunke ([13], p. 134) since H is torsion. By Theorem 2.5, $R(M) = \mathcal{D}_S(M)$ for $S = \{p: Z(p) \notin R\}$ and all cotorsion groups M. Using Proposition 2.2, we see that the correspondence $A \to \mathcal{D}_S(A)$ also defines a cotorsion functor. But if two cotorsion functors agree on the cotorsion groups, they coincide ([13], Theorem 2.6 (iii)). Hence $R = \mathcal{D}_S$.

For a radical class R represented by
$$0 \to Z \xrightarrow{f} G \to H \to 0$$

We shall now try to find out something about G. Let $f(1) = u$. Then for any abelian group A, we have

$$\begin{aligned}
R(A) &= \{g(1): g \in R(\mathrm{Hom}(Z,A))\} \\
&= \{g(1): g \in \mathrm{Hom}(Z,A) \text{ and there exists } h: G \to A \text{ with } g = hf\} \\
&= \{h(u): h \in \mathrm{Hom}(G,A)\}
\end{aligned}$$

In particular, A is in R if and only if for every $a \in A$ there is a map $h \in \mathrm{Hom}(G,A)$ such that $h(u) = a$. But G is in R, so every $a \in G$ has the form $h(u)$ for some $h \in \mathrm{Hom}(G,G)$. Thus $G = \mathrm{End}(G)u$ is a cyclic left $\mathrm{End}(G)$-module.

Let A be in R such that $\mathrm{Hom}(H,A) = 0$. If $a \in A$ then, as above, $a = h(u)$ for some $h \in \mathrm{Hom}(G,A)$. Suppose also $a = k(u)$. Then $(h - k)(u) = 0$; hence $h - k$ factors through $H \cong G/\langle u \rangle$, and so $h = k$. Thus there is a unique $h_a \in \mathrm{Hom}(G,A)$ such that $h_a(u) = a$. For $r \in \mathrm{End}(G)$, $a \in A$, we define

$$ar = h_a r(u)$$

Then $a1 = h_a(u) = a$; $a(r + s) = h_a(r + s)u = h_a(r(u) + s(u)) = h_a r(u) + h_a s(u) = ar + as$. Now $(h_a + h_b)(u) = h_a(u) + h_b(u) = a + b$, so $h_a + h_b = h_{a+b}$. Hence $(a + b)r = (h_a + h_b)r(u) = h_a r(u) + h_b r(u) = ar + br$. Finally, $h_a s(u) = as$, so $h_a s = h_{as}$, and thus $(as)r = h_{as} r(u) = h_a sr(u) = a(sr)$. This proves that A is a right $\mathrm{End}(G)$-module. If $r \in \mathrm{End}(G)$ is such that $r(u) = 0$, then for any $a \in A$, we have $ar = h_a r(u) = 0$. Thus $(0:u) \subseteq (A:0)$.

Now in general the groups A for which $\mathrm{Hom}(H,A) = 0$ are those for which $L(H)(A) = 0$. Since G is in R, it follows that H is in R and hence $L(H) \subseteq R$. Thus $L(H)(A)$ and $A/L(H)(A)$ are in R, and the latter has the right $\mathrm{End}(G)$-module structure described above.

Now let M be a right $\mathrm{End}(G)$-module such that $(0:u) \subseteq (M:0)$. Let R be the ring $\mathrm{End}(G)/(M:0)$. Then (as R-modules) there is an epimorphism $\oplus R \twoheadrightarrow M$. Since $(0:u) \subseteq (M:0)$, there is a group homomorphism $G \cong \mathrm{End}(G)/(0:u) \twoheadrightarrow R$.

Putting these together, we get a group epimorphism $\oplus G \twoheadrightarrow M$, whence M is in R. Since R is closed under extensions, if M is as described and

$$0 \to X \to Y \to M \to 0$$

is an exact sequence of groups with $X \in L(H)$, then Y must be in R.

The foregoing is summarized in our next Theorem.

THEOREM 2.7. Let the radical class R of abelian groups be represented by the sequence $0 \to Z \to G \to H \to 0$. Then

(a) G is cyclic as an End(G)-module

(b) R consists of those groups A for which there is an exact sequence $0 \to X \to A \to Y \to 0$ where $X \in L(H)$ and Y is a right End(G)-module.

The description just given parallels that of D_S given in (1) of §1. The analogue of (2) of §1 is the content of the following result.

THEOREM 2.8. If the radical class R of abelian groups is represented by the sequence $E : 0 \to Z \to G \to H \to 0$, then R is the class of relative injectives for E.

Proof.   A group A is in R if and only if in the induced sequence

$$\text{Hom}(H,A) \to \text{Hom}(G,A) \to \text{Hom}(Z,A)$$

the map on the right is surjective.

For *any* short exact sequence F, we denote by $I(F)$ the class of relative injectives for F and by $E(F)$ the class of short exact sequences relative to which all groups in $I(F)$ are injective. We are then interested in determining for which sequences F starting with Z, $I(F)$ is a radical class. It follows from Propositions 1.1 and 1.2 and Theorem 3.3 of [14] that $I(F)$ is always closed under homomorphic images, direct sums and direct products. The missing property is closure under extensions. When a sequence *does* represent a radical class, we can make one further assertion about the resulting relative homological algebra.

PROPOSITION 2.9. Let the radical class R of abelian groups be represented by a sequence $E : 0 \to Z \to G \to H \to 0$. Then there are enough $E(E)$-injectives.

Proof.   Let $A^{(I)}$ be the direct sum of $|I|$ copies of a group A. For any set I, it is easy to see that every group in $R = I(E)$ is injective relative to

$$0 \to Z^{(I)} \to G^{(I)} \to H^{(I)} \to 0 \qquad\qquad (*)$$

For a group M, let $0 \to K \to Z^{(I)} \to M \to 0$ be a free presentation. Now $(*)$ is in $E(E)$, and a routine argument shows that the induced sequence

$$0 \to Z^{(I)}/K \to G^{(I)}/K \to H^{(I)} \to 0$$

(where $M \cong Z^{(I)}/K$) is also. But since E represents R, we have $G \in R$, whence $G^{(I)}/K \in R$.

One problem with the representation in Theorem 2.7 is that L(H) may coincide with R, in which case we do not gain much. Theorem 2.7 might most usefully be employed in cases where H is a group which belongs to "most" radical classes; torsion H having been disposed of, divisible H seems worth trying next.

A result of O'Donnell ([14], Theorem 7.5) implies that (in the notation of Theorem 2.7) if $A \in R$ is such that $\text{Hom}(H,A) = 0$ then for any abelian group X we have $\text{Hom}(X,A) \in R$.

3.   SOME REMARKS ON MODULES

Most of the results of §2 can be modified so as to apply to modules over any ring R. Some adjustments would need to be made to the cardinal numbers in Theorem 2.1 in the case of large R $(|R| > \alpha)$. In the same Theorem, we need the condition $\cap\{(0:m): m \in M \in R\} = 0$, in order that G should contain an isomorphic copy of R. In the contrary case, since the afore-mentioned intersection is a two-sided ideal I of R, we can treat R, in a natural way, as a radical class of R/I-modules. Thus we can state

THEOREM 3.1.   Let R be a radical class of modules over a ring R. If R is bounded by some infinite cardinal number and is closed under direct products, and if

$$\cap\{0:m): m \in M \in R\} = 0$$

then there is an exact sequence $E : 0 \to R \to G \to H \to 0$ of R-modules such that G is in $\mathcal{R}$ and for every module M we have

$$\mathcal{R}(M) = \mathrm{Im}(\mathrm{Hom}_R(G,M) \to \mathrm{Hom}_R(R,M) \cong M)$$

(the map being induced by E). In particular, $\mathcal{R}$ is the class of relative injectives for E.

COROLLARY 3.2. Let R be any ring. Any radical class $\mathcal{R}$ of R-modules as described in Theorem 3.1 contains all injective R-modules.

The last result is a sort of generalization of the abelian group result to the effect that radical classes containing torsion-free groups must contain all divisible groups. This result is not generally true for modules without the condition

$$\cap\{(0:m): m \in M \in \mathcal{R}\} = 0$$

For example, consider the radical class of all modules M over the ring of integers modulo 6 such that 2M = 0.

REFERENCES

1. S. E. Dickson, On torsion classes of abelian groups, J. Math. Soc. Japan 17 (1965), 30–35.

2. D. W. Dubois, Cohesive groups and p-adic integers, Publ. Math. Debrecen 12 (1965), 51–58.

3. L. Fuchs, Infinite abelian groups, Academic Press, New York and London, 1970 (Vol. I) and 1973 (Vol. II).

4. B. J. Gardner, Torsion classes and pure subgroups, Pacific J. Math. 33 (1970), 109–116.

5. B. J. Gardner, Some closure properties for torsion classes of abelian groups, Pacific J. Math. 42 (1972), 45–61.

6. B. J. Gardner, Two notes on radicals of abelian groups, Comment. Math. Univ. Carolinae 13 (1972), 419–430.

7. R. Göbel, On stout and slender groups, J. Algebra 35 (1975), 39–55.

8. P. A. Griffith, Infinite abelian group theory, University of Chicago Press, Chicago and London, 1970.

9. P. J. Hilton and S. M. Yahya, Unique divisibility in abelian groups, Acta Math. Acad. Sci. Hungar. 14 (1963), 229–239.

10.   A. Hulanicki, A structure of the factor group of an unrestricted sum by
      the restricted sum of abelian groups, Bull. Acad. Polon. Sci. Sér. Sci.
      Math. Astronom. Phys. 10 (1962), 77-80.

11.   A. M. Ivanov, Radikaly v kategorii vsekh abelevykh grupp bez krucheniya,
      Sibirsk. Mat. Zhurnal 20 (1979), 548-556.

12.   A. P. Mišina and L. A. Skornjakov, Abelian groups and modules, Amer.
      Math. Soc. Transl. (2) Vol. 107 (1976).

13.   R. J. Nunke, Purity and subfunctors of the identity, pp. 121-171 of
      Topics In Abelian Groups, Scott, Foresman and Company, Chicago, 1963.

14.   S. R. O'Donnell, A general injectivity for modules, Math. Z. 124 (1972),
      9-19.

15.   E. Sasiada, Proof that every countable and reduced torsion-free abelian
      group is slender, Bull. Acad. Polon. Sci. Sér. Sci. Math. Astronom.
      Phys. 7 (1959), 143-144.

16.   R. Wiegandt, Radical and semisimple classes of rings, Queen's
      University, Kingston, Ontario, 1974.

ON ISÉKI'S BCK-ALGEBRAS

WILLIAM H. CORNISH

School of Mathematical Sciences
The Flinders University of South Australia
Bedford Park, South Australia
Australia

1. INTRODUCTION

BCK-algebras were introduced as an algebraic formulation of C.A. Meredith's
BCK-implicational calculus by Kiyoshi Iséki in 1966 [14]. They form a quasivariety of algebras amongst whose subclasses can be found the earlier implicational models of Henkin [12], algebras of sets closed under set-subtraction,
and dual relatively pseudocomplemented upper semilattices. Many of the
articles in the Mathematics Seminar Notes of Kobe University, Volume 3 (1975)
onwards, are devoted to these algebras; Iséki and Tanaka have provided many
examples and the fundamental first-order theory in [18,19,20], while Iséki's
survey contains many references [15]. Romanowska and Traczyk have done much
to elucidate the nature of the so-called commutative BCK-algebras [32,34].

  Here we review some of the author's recent work, present some new
results and examples, and take the opportunity to tidy a few loose ends.
Section 2 deals with the basics and Section 3 presents examples which are
used throughout the paper. In Section 4, we show that Yutani's quasicommutative varieties are all congruence-3-distributive and so complete an investi-
gation which began in [3]. From Section 5 onwards, we are concerned with

commutative BCK-algebras and Section 6 sees the determination of all varieties of such algebras which satisfy $((xy)y)y = (xy)y$. In Section 7 we show that the variety of bounded commutative BCK-algebras has a proper class of subdirectly irreducibles. Finally, Section 8 presents a representation theorem for finite commutative algebras satisfying $(xy) \wedge (yx) = 0$.

We would like to take this opportunity to thank Professor Iséki for his kind support and encouragement.

2. FUNDAMENTAL IDENTITIES

Let $(A;0)$ be a groupoid with a distinguished element 0; the multiplication is denoted by juxtaposition. On the underlying set, a derived binary relation is defined by

(2.1) $x \leq y$ if and only if $xy = 0$.

Then, $(A;0)$ is a *BCK-algebra* if it satisfies the following identities:

(2.2) $(xy)(xz) \leq zy$

(2.3) $x(xy) \leq y$

(2.4) $x \leq x$

(2.5) $0 \leq x$

and the *quasi-identity*

(2.6) If $x \leq y$ and $y \leq x$, then $x = y$

Notice that 0 is defined equationally; (2.5) says $xx = yy = 0$. In effect, BCK-algebras form a quasivariety of groupoids. It is not known whether they form a variety.

The following two quasi-identities are easy to obtain:

(2.7) $z \leq y$ implies $xy \leq xz$

(2.8) $x \leq y$ and $y \leq z$ imply $x \leq z$

Thus $(A;\leq)$ is a partially ordered set with 0 as its smallest element. Let $\bar{A}$ denote the corresponding order-dual. For $x \in A$, consider $t_x: \bar{A} \to \bar{A}$, $t_x(y) = xy$, for each $y \in A$. Due to (2.3), $t_x t_x(y)$ always exceeds y in $\bar{A}$. Due to (2.7), the pair (t_x, t_x) is a Galois connection between $\bar{A}$ and itself. Hence, $t_x^3 = t_x$ and t_x takes suprema in $\bar{A}$ into infima in $\bar{A}$. That is, for any x,y and z in A, $x(x(xy)) = xy$, and when $y \wedge z$ exists, $xy \vee xz$ exists and $x(y \wedge z) = xy \vee xz$. Hence, Galois connections provide us with a neat summary of these early consequences of the BCK-axioms.

Another consequence is the following crucial identity; it follows from a skillful use of (2.2), see Iséki and Tanaka [18 or 20].

(2.9)$(xy)z = (xz)y$

Then the following quasi-identities are easily obtained:

(2.10)$xy \leq x$

(2.11)$x0 = x$

(2.12)$xy \leq z$ if and only if $xz \leq y$

(2.13)$z \leq y$ implies $zx \leq yx$

The underlying order is of fundamental importance despite the fact that a BCK-algebra is not an ordered groupoid in the usual sense, as seen by the behavior of the multiplication in (2.7) and (2.13).

We define the groupoid-polynomials xy^n inductively by:

$$xy^0 = x, \quad xy^{k+1} = (xy^k)y \text{ for } k \geq 0$$

Due to (2.10), we have the descending chain of polynomials

(2.14)$\quad x \geq xy \geq xy^2 \geq \ldots \geq xy^{k+2} \geq \ldots$

Using these polynomials it is possible to obtain new identities which are fundamental. For example, $(xy^n)(xz) \leq zy^n$, and $(xy^n)z^m = (xz^m)y^n$. Also, we shall be concerned with the following identity:

$(E_n) \quad xy^n = xy^{n+1}$

It is equivalent to either of $(xy^n)y^n = xy^n$ and $(xy)z^n = (xz^n)(yz^n)$, when $n \geq 1$. Of course, (E_0) defines the trivial variety. The proof that the second identity is a consequence of (E_n) is tricky; it is along the lines of the proof of Theorem 8 of Iséki and Tanaka [20].

LEMMA 2.1. ([6; Lemma 1.3]) If a BCK-algebra satisfies the identity (E_n), then it also satisfies

$(C_n) \quad (x(xy)^n)(yx)^n = (y(yx)^n)(xy)^n$

In [36], Yutani defined the polynomials $q_{m,n}(x,y)$ by double induction as follows:

$$q_{0,0}(x,y) = x(xy), \quad q_{m+1,n}(x,y) = q_{m,n}(x,y)(xy), \quad q_{m,n+1}(x,y)$$

$$= q_{m,n}(x,y)(yx)$$

They are well-defined because of (2.9). He said that a BCK-algebra is *quasicommutative* when there exist i,j,m,n such that the algebra satisfies the identity $q_{i,j}(x,y) = q_{m,n}(y,x)$. This is easier to understand if we use the polynomials xy^n. Indeed, induction shows that $q_{m,n}(x,y) = (x(xy)^{m+1})(yx)^n$. Thus, we see that a BCK-algebra is quasicommutative when it satisfies the identity:

$$(C_{m,n}^{i,j})(x(xy)^i)(yx)^j = (y(yx)^m)(xy)^n$$

for suitable integers $i,m \geq 1$ and $j,n \geq 0$. Notice that when either i or m is 0, we get the trivial variety. Also observe that the identity (C_n) is the identity $(C_{n,n}^{n,n})$, and so if an algebra satisfies (E_n), then it is quasicommutative due to Lemma 2.1.

These identities are very useful when finite algebras are considered. Because of (2.14), we must have, for any ordered pair (a,b) of elements in a finite BCK-algebra A, an integer n(a,b) such that $ab^{n(a,b)} = ab^{n(a,b)+1}$. Put $n = \max\{n(a,b) : (a,b) \in A \times A\}$. Then, A satisfies the identity (E_n), and so the identity $(C_n) = (C_{n,n}^{n,n})$. In [16], Iséki also observed that a finite BCK-algebra must satisfy a suitable identity (E_n); see also Iséki [17]. Thus, *a finite BCK-algebra is quasicommutative;* this was first shown by Yutani [36], where, in the proof of his Theorem 2, he proved that a finite algebra must satisfy a suitable identity $(C_{m,n}^{m,n})$.

Let $\underline{E}_n$, $\underline{C}_n$ and $\underline{C}_{m,n}^{i,j}$ denote the classes of all BCK-algebras which satisfy (E_n), (C_n) and $(C_{m,n}^{i,j})$, respectively. The first part of the following result is due to Yutani [36; Theorem 1], while the second part is Theorem 1.4 of [6] and it follows from Lemma 2.1.

THEOREM 2.2.   The class $\underset{=m,n}{C^{i,j}}$ is a variety of type $\langle 2,0 \rangle$ with

(1) $((xy)(xz))(zy) = 0$

(2) $0x = 0$

(3) $x0 = x$

and

(4) $(C_{m,n}^{i,j})$

as an equational base.

The class $\underset{=n}{E}$ is a variety; its equational base is provided by (1), (2), (3), (C_n), and (E_n).

It follows that a finite BCK-algebra generates a variety of BCK-algebras.

3.   EXAMPLES

3.1.   Let $(A;\leq,0)$ be a partially ordered set with a smallest element 0. Define

$$xy = \begin{cases} 0 \text{ if } x \leq y \\ \\ x \text{ if } x \nleq y \end{cases}$$

Then we get a BCK-algebra in which the original order is consistent with (2.1). This algebra satisfies (E_1).

Algebras satisfying (E_1) are generally called *positive implicative* BCK-algebras. They are precisely the *implicative models* of Henkin [12]. Henkin's axioms are

(1) $xy \leq x$

(2) $(xz)(yz) \leq (xy)z$

(3) $0 \leq x$

and

(4) $x \leq y$ and $y \leq x$ imply $x = y$, where 0 is nullary and $\leq$ is defined by (2.1).

3.2 Let $(A; \vee)$ be a dual relatively pseudocomplemented upper semilattice, that is, for any $x, y \in A$, there is a (unique) smallest element xy in $\{z \in A: x \le y \vee z\}$. Thus, $x \le y \vee z$ if and only if $xy \le z$. Then $0 = xx$ is the smallest element of A for any $x \in A$, and (2.1) holds. Better still, A becomes a positive implicative BCK-algebra. For details, see [5].

3.3 Let A be a non-empty subset of a Boolean algebra $(B; \wedge, \vee, ', 0, 1)$ which is closed under subtraction. On A, or B, define $xy = x \backslash y = x \wedge y'$. Then A and B become BCK-algebras which satisfy the identity

(I) $x(yx) = x$

Algebras satisfying (I) are called *implicative* BCK-algebras, and always arise as reducts of Boolean algebras in the above manner. This was first proved by Kalman in 1960. Other proofs have been given by Abbott, Shafaat, Thaheem, Iséki and Tanaka. For a history, see the author's paper [1].

The class $\underline{\underline{I}}$ of implicative BCK-algebras is a variety. It is the variety generated by the two-element BCK-algebra $A_1 = \{0, a_1: 0 < a_1; a_1 a_1 = 0 a_1 = 00 = 0; a_1 0 = a_1\}$, which is a subalgebra of any non-trivial BCK-algebra. Hence, $\underline{\underline{I}}$ *is the smallest non-trivial (quasi-) variety of BCK-algebras.* In effect, this result goes back to Kalman [24].

Notice that any implicative BCK-algebra satisfies (E_1) and also $(C_{1,0}^{1,0})$, that is

(T) $x(xy) = y(yx)$

Indeed, both sides represent the infimum. Algebras satisfying (T) are called *commutative* BCK-algebras. They were first considered by Tanaka. Theorem 9 of [20] says that $\underline{\underline{I}} = \underline{\underline{T}} \cap \underline{\underline{E}}_1$, where $\underline{\underline{T}}$ is the variety of commutative BCK-algebras.

3.4 Let A be a ring with central covers. That is, for each $x \in A$, there is a central idempotent $C(x)$ such that

(1) $xC(x) = x$

(2) when e is a central idempotent, $xe = x$ implies $C(x)e = C(x)$

Examples abound and include finite rings, Boolean rings, Baer rings in the sense of Kaplansky, and Rickart rings with no nilpotent elements. If the ring-product is denoted by $x.y$, then A has an implicative BCK-reduct when the BCK-product is defined by $xy = x.C(x-y)$. For more details, see [2].

Notice that when A is a Boolean ring with an identity element,

$$xy = x.(x - y) = x.(x + y) = x + xy = x.(1 + y) = x \wedge y'$$

in the associated Boolean algebra. This is consistent with 3.3.

3.5 Let M be any non-empty set. Let S_M be the tree with smallest element 0, unique atom c, and M as its set of maximal elements; there are no other elements. By defining $m_1 m_2 = c$ for any $m_1, m_2 \in M$, and the other products in a manner consistent with (2.1), (2.4) and (2.5), S_M becomes a BCK-algebra. In fact, it is commutative and satisfies (E_2) also. Moreover, it is simple and so subdirectly irreducible. This example is due to Setō [33].

Thus, $\underline{\underline{T}}$ has a class of (simple) subdirectly irreducible algebras. In fact, due to our remarks in 2.3, $\underline{\underline{T}} \cap \underline{\underline{E}}_1$ is equationally complete, while $\underline{\underline{T}} \cap \underline{\underline{E}}_2$ is *not* residually small.

For another *class* of simple trees in $\underline{\underline{T}} \cap E_{n+1}$, $n > 1$, see Example 5 of Iséki and Tanaka [19].

3.6 *Abelian ℓ-group cones*. Let G be a lattice-ordered group and $A = G^+$ its positive cone. On A, define $xy = (x - y)^+ = (x - y) \vee 0$. Then A becomes a BCK- algebra, whose original order is consistent with (2.1) if and only if G is abelian. When G is abelian, A is a commutative BCK-algebra and such algebras have been abstractly characterized by the author in [4].

When G is the o-group Z of integers, we get a simple commutative BCK-algebra, which has been studied at length by Halkowska [11]. Here we are relying on Math. Reviews.

If A is any BCK-algebra and $x \in A$, then $(x] = \{y \in A: y \leq x\}$ is a sub-algebra. Thus, $A = Z^+$ gives us many important algebras. We use the fol-lowing notation:

Let $0 = a_0 < a_1 < ..... < a_n < .....$ be a chain of order type ω. From the above comments, it becomes a commutative BCK-algebra A_ω, when $a_n a_m$ is defined to be $a_{\max(n-m,0)}$. The subalgebra $(a_k]$ is denoted by A_k. It is not difficult to see that $A_k \in \underline{\underline{T}} \cap \underline{\underline{E}}_k$, A_k is simple, and $A_k \notin \underline{\underline{E}}_{k-1}$, for any $k \geq 1$. Of course, $\underline{\underline{E}}_k \subseteq \underline{\underline{E}}_{k+1}$. *Hence, the varieties* $\underline{\underline{E}}_n$ *and* $\underline{\underline{T}} \cap \underline{\underline{E}}_n$ *both form strictly increasing chains.*

We will return to the construction of 3.6 later. We now turn to the behaviour of congruences.

4. CONGRUENCES

Let Φ be a congruence on a BCK–algebra A and ker $\Phi = \{x \in A: x \equiv 0(\Phi)\}$.
Then, $K = $ ker Φ has the following two properties:

(1) $0 \in K$;

(2) for any $x,y \in A$, $x \in K$ whenever $xy, y \in K$.

Such a subset is called an *ideal*. Conversely, given any ideal J, we get a
congruence $\Theta(J)$, given by $x \equiv y(\Theta(J))$ if and only if $xy, yx \in J$. By Theorem
2 of Iséki and Tanaka, the associated quotient algebra is a BCK–algebra [19].

Now suppose A is in a variety of BCK–algebras. Then the quotient
algebra A/Φ is a BCK–algebra and so (2.6) applies to it. Hence, for any
$x,y \in A$, $x \equiv y(\Theta(\text{ker } \Phi))$ iff $xy, yx \in$ ker Φ, iff $xy, yx \equiv (\Phi)$ iff $\bar{x} \leq \bar{y}$ and
$\bar{y} \leq \bar{x}$ iff $\bar{x} = \bar{y}$ iff $x \equiv y(\Phi)$, where $\bar{x}$ and $\bar{y}$ are the Φ–classes of x and y
respectively. In other words, for any algebra A in a variety of BCK–
algebras, the maps $K \to \Theta(K)$ and $\Phi \to$ ker Φ are mutually inverse lattice-
isomorphisms between the lattice, $J(A)$, of ideals of A and the lattice,
$\text{Con}(A)$, of congruences on A.

Let B be a non–empty subset of a BCK–algebra A. Due to Theorem 3 of
Iséki and Tanaka [19], <B>, the ideal generated by B, is given by

$$\langle B \rangle = \{a \in A: (\ldots(ab_1)\ldots)b_t = 0 \text{ for some } b_1,\ldots,b_t \in B\}$$

When $B = \{b_1,\ldots,b_r\}$, we use $\langle b_1,\ldots,b_r \rangle$ to denote $\langle B \rangle$; (2.5) and (2.9) allow
us to say that $\langle b_1,\ldots,b_r \rangle = \{a \in A: (\ldots(ab_1^{n_1})\ldots)b_r^{n_r} \text{ for some}$
$n_1,\ldots,n_r \geq 0\}$.

Let $\Theta(a,b)$ be the smallest congruence which identifies a and b. Then
combining our observations, we can assert:

THEOREM 4.1. Any variety $\underline{V}$ of BCK–algebras has the Congruence Extension
Property. Moreover, for any algebra A in $\underline{V}$, and $x,y,a,b \in A$, $x \equiv y(\Theta(a,b))$
if and only if there exists an integer $k \geq 1$ such that $((xy)(ab)^k)(ba)^k = $
$0 = ((yx)(ab^k)(ba)^k$.

Proof.   Let A_1 be a subalgebra of $A \in \underline{V}$ and $K_1 \in J(A_1)$. Because of
the description of $\langle K_1 \rangle$, the ideal of A generated by K_1, $K_1 = \langle K_1 \rangle \cap A_1$.
The Congruence Extension Property now follows from the correspondence between
ideals and congruences.

As for the second assertion, $x \equiv y\Theta(a,b)$ iff $xy,yx \in \langle ab,ba \rangle$ iff there are integers r,s,t,u such that $((xy)(ab)^r)(ba)^s = 0 = ((yx)(ab)^t)(ba)^n$ iff $((xy)(ab)^k)(ba)^k = 0 = ((yx)(ab)^k)(ba)^k$, where k is the maximum of r,s,t and u.

For some recent results about the Congruence Extension Property on a finitary algebra, see Fried [9].

COROLLARY 4.2. ([6; Theorems 2.4, 2.5]) Let A be an algebra in the variety $\underline{\underline{E}}_n$ and $x,y,a,b \in A$. Then $x \equiv y(\Theta(a,b))$ if and only if $((xy)(ab)^n)(ba)^n = 0 = ((yx)(ab)^n)(ba)^n$.

Moreover, when $H = \langle a_1,\ldots,a_t \rangle$, $K = \langle b_1,\ldots,b_r \rangle$ are two finitely generated ideals of A, and for $i = 1,\ldots,t$, $d_i = (\ldots(a_i b^n)\ldots)b_r^n$, the dual relative pseudocomplement HK (c.f. 2.2) of H and K in the upper semilattice of finitely generated ideals is the ideal $\langle d_1,\ldots,d_t \rangle$.

The two statements of 4.2 are connected because of a recent theorem of Köhler and Pigozzi [25], see [6] for a discussion. As mentioned in [6], 4.2 implies that the ideal-lattice $J(A)$, and so the congruence-lattice $\mathrm{Con}(A)$, is distributive for any $A \in \underline{\underline{E}}_n$. However, more can be obtained by using "Mal'cev conditions".

A finitary algebra A is said to be *n-permutable*, for some given integer $n \geq 2$, if the n-fold alternating relational products $\Theta\Phi \ldots$ and $\Phi\Theta \ldots$ are equal, for any two congruences Θ and Φ on A. This concept is a generalization of permutability (= 2-permutability). Hagemann and Mitschke characterized an n-permutable variety, that is a variety with all its members n-permutable, in terms of the existence of $n - 1$ ternary polynomials satisfying certain identities. In particular, a variety is 3-permutable if and only if there are two ternary polynomials $r(x,y,z)$ and $s(x,y,z)$ such that each algebra in the variety satisfies the identities $r(x,z,z) = x$, $s(x,x,z) = z$ and $r(x,x,z) = s(x,z,z)$; compare this with Mal'cev's classical result which says that a variety is permutable if and only if there is a ternary polynomial $p(x,y,z)$ such that each member of the variety satisfies $p(x,z,z) = x = p(z,z,x)$. On the other hand, a variety is *congruence-distributive* if the lattice of congruences of each of its algebras is distributive. Jónsson has shown that a variety is congruence-distributive if and only if it is *congruence-n-distributive*, or more briefly *n-distributive*, in the sense that there exists an integer $n \geq 2$ and $n - 1$ ternary polynomials satisfying certain identities. For example, a variety is 2-distributive if and only if there is a polynomial $m(x,y,z)$ such that $m(x,x,y) = m(x,y,x) = m(y,x,x) = x$ on each member of the variety. More importantly for us, a variety is 3-distributive if and only if there exist polynomials $t_1(x,y,z)$, $t_2(x,y,z)$ such that each algebra in the variety satisfies:

(1) $t_1(x,y,x) = x = t_2(x,y,x)$

(2) $t_1(x,x,z) = x$

(3) $t_2(x,x,z) = z$

and

(4) $t_1(x,z,z) = t_2(x,z,z)$

Proofs of these results on n-permutability and n-distributivity can be found in Jónsson's article [22].

These conditions are connected. In [3; Theorem 2.6], the author showed that an n-permutable variety is congruence-distributive if and only if it is n-distributive. This is an immediate consequence of [3; Theorem 2.5] which says that a variety is n-distributive if and only if, for any congruences Θ, Φ_1 and Φ_2 on an algebra in the variety, $\Theta \cap (\Phi_1 \Phi_2)$ is contained in the n-fold alternating product $(\Theta \cap \Phi_1)(\Theta \cap \Phi_2) \ldots$ When it comes to the value of n, the cases n = 2 and n = 3 seem to be the most significant. 3-permutability implies congruence-modularity; 3-permutability and 3-distributivity combine to give a nice refinement of the Second Isomorphism Theorem, see [3; Theorem 2.2]. These conditions also restrict the size of an equational base. Indeed, let $\underline{V}$ be a finitely based variety. Then Padmanabhan and Quackenbush have shown that $\underline{V}$ is n-based if it is n-distributive [30]. Also they proved that $\underline{V}$ is 1-based if it is 2-permutable and 2-distributive. In [7], the author recently exhibited a 2-based variety which is 3-permutable and 3-distributive but neither 1-based nor 2-permutable nor 2-distributive. The example is the variety of groupoids which satisfies the identities $x(yx) = x$ and $x(xy) = y(yx)$; it is closely related to the variety $\underline{\underline{I}}$ of 3.3.

The next result tidies up the author's work on the relevance of these conditions for BCK-varieties.

THEOREM 4.3. No non-trivial variety of BCK-algebras is either permutable or 2-distributive. However, any quasicommutative $C_{m,n}^{i,j}$, as well as any variety $\underline{\underline{E}}_n$ is 3-permutable, 3-distributive and 3-based.

With the preceding notation, the polynomials which ensure the 3-permutability of $\underset{=m,n}{C^{i,j}}$ are

$$r(x,y,z) = (x(yz)^i)(zy)^j \text{ and } s(x,y,z) = (z(yx)^m)(xy)^n$$

while those which ensure 3-distributivity are

$$t_1(x,y,z) = (x((xy)(zy))^i)((yx)(yz))^j$$

and

$$t_2(x,y,z) = (z((yx)(yz))^m)((xy)(zy))^n$$

Proof. The first statement is Corollary 3.16 of [3]. It holds for any non-trivial variety as it holds in the smallest such variety $\underline{\underline{I}}$, see the remarks in 3.3. Secondly, the $\underline{\underline{I}}$-algebra obtained, as in 3.3, by removing the largest element from the 8-element Boolean algebra provides a suitable counterexample within $\underline{\underline{I}}$. For the real reasons behind this counterexample, see [3; Theorems 3.14, 3.15].

The remainder follows from our preceding remarks and a routine verification. When it comes to $\underline{\underline{E}}_n$, recall that Theorem 2.2 says that $\underline{\underline{E}}_n$ is a subvariety of $\underline{\underline{C}}_n = C_{n,n}^{n,n}$.

COROLLARY 4.4. ([6; Theorem 2.2]) Any finite BCK-algebra generates a variety of BCK-algebras, which is 3-permutable, 3-distributive and 3-based.

Proof. This follows from the remarks towards the end of Section 2, Theorem 4.3, and Baker's Theorem which says that a finite algebra in a congruence-distributive variety of finite similarity type generates a finitely based subvariety. For a proof and discussion, see [22; §62, pp. 373-377].

5. COMMUTATIVE BCK-ALGEBRAS

Recall from 3.2 that a BCK-algebra is said to be *commutative* when it satisfies the identity:

(T) $x(xy) = y(yx)$

Commutative BCK-algebras form a very interesting class. Due to Theorem 2.2, it is a variety. Yutani [35] showed that this variety $\underline{\underline{T}}$ has the following identities as its equational base:

(1)   xx = 0

(2)   x0 = x

(3)   x(xy) = y(yx)

and

(4)   (xy)z = (xz)y

At the end of [1], it was observed that these identities are independent.

When $(A;0)$ is a commutative BCK-algebra, the underlying partially ordered set $(A;\leq)$ is a lower semilattice. Moreover, the infimum is the derived operation $x \wedge y = x(xy) = y(yx)$. This fundamental fact is due to Tanaka; see either [18] or [20; Theorem 3] for a proof. Now let $x \in A$ be fixed and consider the principal order-ideal $(x] = \{y \in A: y \leq x\}$. It is a subalgebra of $(A;0)$, and also, for any $y \in (x]$, $N_x N_x y = y$, where $N_x: (x] \to (x]$ and $N_x y = xy$. Due to (2.7), N_x is then an involution (dual-order-isomorphism which is its own inverse) on the lower semilattice $(x]$. Hence, $(x]$ is a lattice, wherein the supremum of $y, z \in (x]$ is given by $x \vee y = N_x(N_x y \wedge N_x z)$. This means that $(A;\leq)$ is a *nearlattice* in the sense that $(A;\leq)$ is a lower semilattice and each principal order-ideal is a lattice, or equivalently, $(A;\leq)$ is a lower semilattice with the *upper bound property* in that any two elements of A have a least upper bound whenever they share a common upper bound. Nearlattices have been studied in detail by the author and his students Robert Hickman and Abu Noor; see Cornish and Hickman [8], Hickman [13], and Noor [29]. They are used in [3; Section 3] to study commutative BCK-algebras; it was this connection which stimulated the author's interest in BCK-algebras. The upper bound property in commutative BCK-algebras was also observed by Traczyk [34; Lemma 2.1]. Iséki and Tanaka [18, 20, Theorem 6] had already shown that a bounded commutative BCK-algebra is a lattice with an involution; a BCK-algebra $(A;0)$ is called bounded if the underlying partially ordered set $(A;\leq)$ has a largest element 1, or equivalently, there is an element $1 \in A$ such that

(B)   x1 = 0 for all $x \in A$

Actually, the underlying lattice is distributive. This was first proved by Traczyk [34]. Another proof was given by the author in [3; Section 3]; this proof is of independent interest and we now outline it.

THEOREM 5.1. Suppose a lattice L, with a smallest element 0, possesses the following two properties.

(1) For any $x,y \in L$ with $y \leq x$, the complement of y in the sublattice $(x]$ is unique, whenever it exists.

(2) For any $x \in L$, the sublattice $(x]$ has an involution N_x

Then L is distributive. Consequently, the underlying partially ordered set of a bounded commutative BCK-algebra is a distributive lattice.

Proof.   The distributivity of L follows by showing that (1) and (2) exclude the pentagon and the diamond. In [20; Proposition 7], Iséki and Tanaka showed that complements are unique, whenever they exist in a bounded commutative BCK-algebra. It follows that (1) and (2) are valid in such an algebra.

Actually, a bounded commutative BCK-algebra is a *Kleene algebra* in that it is a bounded distributive lattice with an involution N and the identity $x \wedge Nx \leq y \vee Ny$ is universally satisfied, [3; Theorem 3.11]. The proof rests upon the following lemma, which we highlight for subsequent developments. It is identity (26) of Traczyk [34], and is not easy to establish; see also Pałasiński for another derivation [31; Lemma, p. 139].

LEMMA 5.2. (Traczyk [34]) A bounded commutative BCK-algebra satisfies the identity

(L) $(xy) \wedge (yx) = 0$

Call an ideal P of a commutative BCK-algebra *prime* if $P \neq A$ and either $a \in P$ or $b \in P$ whenever $a \wedge b \in P$. As ideals are hereditary, $K_1 \cap K_2 = \{k_1 \wedge k_2 : k_1 \in K_1,\ k_2 \in K_2\}$ for any ideals K_1, K_2 in commutative A; in particular, $\langle a \rangle \cap \langle b \rangle = \langle a \wedge b \rangle$ for any $a, b \in A$. Hence, an ideal $P \neq A$ is prime when and only when $K_1 \cap K_2 \subseteq P$ implies $K_1 \subseteq P$ or $K_2 \subseteq P$, for any ideals K_1, K_2 of A. Due to the correspondence between ideals and congruences in a variety and the congruence-distributivity of the variety $\underline{\underline{T}}(= \underline{\underline{C}}_{1,0}^{1,0})$, implied by Theorem 4.3, $\cap \{P: P$ prime ideal in $A\} = \{0\}$ and A is isomorphic to a subdirect product of the quotient algebras $A/\Theta(P)$. The next result incorporates Theorem 3.3 of [6] and Lemma 5.2 of Romanowska and Traczyk [32]. Recall that a *tree* is a lower semilattice in which the least upper bound of two elements exists if and only if these elements are comparable, that is, it is a nearlattice in which each principal order-ideal is a chain.

THEOREM 5.3. Let A be a commutative BCK-algebra. Then an ideal $P \neq A$ is prime if and only if the associated quotient algebra is a tree in which 0 is meet-irreducible. When A satisfies (L), an ideal $P \neq A$ is prime if and only if its associated quotient is a chain.

Hence, A is isomorphic to a subdirect product of algebras, each of which is a tree having 0 as a meet-irreducible element. Moreover, A satisfies (L) if and only if it is isomorphic to a subdirect product of totally ordered algebras.

Proof.   By now the only unclear part should be the assertion about trees. Because (x] has an involution for any x in a commutative BCK-algebra B, 0 is meet-irreducible if and only if x is join-irreducible. This forces (x] to be a chain for any $x \in B$ when 0 is meet-irreducible, and so B is a tree; this elegant argument is due to Romanowska and Traczyk [32].

COROLLARY 5.4. (Pałasiński [31; Theorem 1]) If a commutative BCK-algebra is a lattice, then it is isomorphic to a subdirect product of totally ordered algebras.

Proof.   Suppose A is a lattice, $x,y \in A$ and $z = x \vee y$. Then x, y, xy, $yx \in (z]$, due to (2.10), and (z] is a subalgebra and sublattice of A, which is bounded. Hence, Lemma 5.2 applies to (z] and so A satisfies (L).

The converse of this corollary is false. This is easily seen from 3.3. It should also be mentioned that Pałasiński [31; Theorem 3] showed that a subdirectly irreducible BCK-algebra is totally ordered if and only if it satisfies $(x(x(yz)))(x(zy)) = 0$.

We now turn to the connection with the varieties $\underline{\underline{E}}_n$.

THEOREM 5.5. ([6; Proposition 3.1, Corollary 3.2]) A commutative BCK-algebra satisfies (E_n) if and only if it satisfies $x \wedge (yx^n) = 0$. Hence, a subdirectly irreducible commutative algebra in $\underline{\underline{E}}_n$ is simple.

Proof.   When A is subdirectly irreducible and $x \wedge (yx^n) = 0$, with $x \neq 0$ in A, Theorem 5.3 implies $yx^n = 0$, that is, $y \in \langle x \rangle$. Hence A is simple.

This result is interesting, especially when we recall that 3.5 says that $\underline{\underline{T}} \cap \underline{\underline{E}}_n$ has a class of simple algebras for any $n \geq 2$.

In [26, 27], Komori considered a variety of groupoids which turn out to be the groupoid-duals (opposites) and order-duals of commutative BCK-algebras satisfying (L). In fact, Theorem 3.13 of [27] can be interpreted as the following important result. The proof uses the identity $x \wedge (yx^n) = 0$ of Theorem 5.5 to show firstly that the algebra is atomic and then that the atom generates the algebra; it is quite complex.

LEMMA 5.6.   (Komori [27]) A commutative totally ordered algebra in $\underline{\underline{E}}_n$ is isomorphic to the algebra A_m of 3.6 for some $m \le n$.

Combining our results, we obtain

THEOREM 5.7.   The subvariety of $\underline{\underline{E}}_n$ determined by the identities (T) and (L) is the variety generated by the algebra A_n of 3.6.

In the next section, we use 5.3 and 5.6 to determine the lattice of subvarieties of $\underline{\underline{T}} \cap \underline{\underline{E}}_2$.

6.   THE VARIETY $\underline{\underline{T}} \cap \underline{\underline{E}}_2$

THEOREM 6.1. Apart from the 2-element algebra A_1 of 3.3 and 3.6, an algebra in $\underline{\underline{T}} \cap \underline{\underline{E}}_2$ is subdirectly irreducible (simple) if and only if it is isomorphic to an algebra S_M of 3.5 for some non-empty set M. Hence, up to isomorphism, A_1 and the algebras S_M, with M being finite, are the finitely generated subdirectly irreducibles in $\underline{\underline{T}} \cap \underline{\underline{E}}_2$.

Proof.   Let A be a subdirectly irreducible algebra in $\underline{\underline{T}} \in \underline{\underline{E}}_2$, which has at least three elements. Let $a \in A$. Due to Theorem 5.3, A is a tree and so the subalgebra (a] is totally ordered. By Lemma 5.6, this chain (a] has at most three elements. If b is another element and $a,b \ne 0$ then by Theorem 5.3, $a \wedge b > 0$. Hence, A is order-isomorphic to the underlying semilattice of some suitable S_M. Suppose a,b are maximal elements of A and c is the unique atom of A. As $c \le b$, $ab \le ac$. But due to Lemma 5.6, $ac = c$, so that $ab \le c$. But $ab \ne 0$ as $a \not\le b$. Hence, $ab = c$. It now follows that A is isomorphic to S_M as a BCK-algebra. The last statement of the theorem is obvious.

COROLLARY 6.2.   The variety $\underline{\underline{T}} \cap \underline{\underline{E}}_2$ is locally finite.

Proof.   This follows from the last statement of 6.1. Indeed, we see that the finitely generated subdirectly irreducibles are finite, and for each k, only finitely many are k-generated. Thus each finitely generated algebra is finite, as asserted.

When M is finite with n elements, we use S_n in place of S_M. When M is countably infinite, we use S_ω in place of S_M. Notice that $A_1 \leq \ldots \leq S_n \leq S_{n+1} \leq \ldots \leq S_\omega$, where $\leq$ means "is a subalgebra of". Due to the congruence-distributivity of $\underline{\underline{T}} \cap \underline{\underline{E}}_2$, Jónsson [21; Corollary 3.4] says that there exists an identity which holds in S_n but does not hold in S_{n+1}. We now give such an identity.

THEOREM 6.3.   Let $n \geq 1$ and let $\underline{V}(S_n)$ be the variety generated by S_n. Then an equational base for $\underline{V}(S_n)$ is provided by the equational base for $\underline{\underline{T}} \cap \underline{\underline{E}}_2$, together with the identity

$$(S_n): \bigwedge_{1 \leq i \leq n} (x_i x_{i+1}) \wedge x_{n+1} x_1 = 0$$

Proof.   It is not hard to check that S_m satisfies (S_n) for $1 \leq m \leq n$. Also S_{n+1} fails to satisfy (S_n). The result follows from the last statement of 6.1.

Finally, we give the lattice of subvarieties of $\underline{\underline{T}} \cap \underline{\underline{E}}_2$. It is a chain of type $\omega + 1$. The proof is along the lines of Theorem 4 in Lakser [28], wherein it is shown that the lattice of varieties of distributive pseudocomplemented lattices is a chain of type $\omega + 1$. In fact, Kagan and Quackenbush [23; comments after Theorem 5.3] have pointed out that Lakser's argument depends upon the following properties, which hold for $\underline{\underline{T}} \cap \underline{\underline{E}}_2$:

(1)   Each algebra has a distributive congruence lattice

(2)   The finite subdirectly irreducible algebras form an increasing chain

(3)   Finitely generated algebras are finite

(4)   Each subdirectly irreducible is a 1 - 1 direct limit of finite subdirectly irreducibles.

THEOREM 6.4.   The lattice of subvarieties of $\underline{\underline{T}} \cap \underline{\underline{E}}_2$ is

$$\underline{0} \subset \underline{\underline{I}} \subset \underline{V}(S_1) \subset \ldots \subset \underline{V}(S_n) \subset \ldots \subset \underline{\underline{T}} \cap \underline{\underline{E}} = \underline{V}(S_\omega)$$

In connection with 6.4, $\underline{0}$ is the trivial variety with base x = y, $\underline{\underline{I}}$ = $\underline{\underline{V}}(A_1)$ as in 3.3, $S_1 = A_2$ and (S_1) is (L) of 5.2.

7. BOUNDED COMMUTATIVE BCK-ALGEBRAS

From now on, we regard a bounded commutative BCK-algebra as an algebra (A;0,1) of type (2,0,0). Such algebras form a variety $\underline{\underline{T}}'$ whose equational base consists of the identities in a base for $\underline{\underline{T}}$, together with (B): x1 = 0, of Section 5, regarded as an identity.

Not only do $\underline{\underline{T}}'$-algebras have a derived infimum x ∧ y = x(xy), they also have a derived supremum x ∨ y = 1(1x) ∧ (1y).

THEOREM 7.1. ([3; Theorem 1.10]) The variety $\underline{\underline{T}}'$ is permutable, 2-distributive and 1-based.

The polynomial p(x,y,z) = (x(yz)) ∨ (z(yx)) satisfies p(x,x,z) = p(z,x,x) = z on $\underline{\underline{T}}'$ and so ensures permutability. The polynomial m(x,y,z) = (x(xy)) ∨ (y(yz)) ∨ (z(zx)) satisfies m(x,x,z) = m(x,z,x) = m(z,x,x) = x on $\underline{\underline{T}}'$ and so ensures 2-distributivity.

The comment on "1-based" follows from the other two properties and Theorem 2 of Padmanabhan and Quackenbush ⌊30⌋. Using the congruence properties of 7.1, the fact that a finite algebra is within some variety $\underline{\underline{E}}_n$, Lemma 5.2 and Lemma 5.6, we easily get

THEOREM 7.2. (Romanowska and Traczyk [32], [6; Theorem 3.7]) A finite bounded commutative BCK-algebra is isomorphic to the direct product of simple totally ordered BCK-algebras. Consequently, its congruence-lattice is a Boolean lattice.

The finite totally ordered commutative BCK-algebra A_n of 3.6 gives rise to a $\underline{\underline{T}}'$-algebra, which is denoted by A_n^1. Using 7.1, it is not hard to see that A_n^1 is quasiprimal, in fact semiprimal. Moreover, a representation of the ternary discriminator is known. Also there is a description of the variety $\underline{\underline{T}}^1 \cap \underline{\underline{E}}_n$. We will not repeat the details here, but rather we refer to [6; Theorem 3.8]. Instead, we turn to a way of producing subdirectly irreducible (but not simple) algebras in $\underline{\underline{T}}'$.

Let our notation be as in 3.6, so that A is a commutative BCK-algebra which is the positive cone of an abelian ℓ-group G and $xy = (x - y)^+$ for any $x,y \in A$.

Let $\bar{A}$ denote the order-dual of A. For $a \in A$, let $\bar{a}$ denote the corresponding element of $\bar{A}$. Then $\bar{A} = \{\bar{a}: a \in A\}$ and $\bar{a} \le \bar{b}$ in $\bar{A}$ if and only if $b \le a$ in A.

Let B be the ordinal sum $B = A \oplus \bar{A}$. We define a product on B as follows: for any $x,y \in A$, let:

$$xy = (x - y)^+, \text{ as before}$$

$$\bar{x}\bar{y} = \overline{yx}$$

$$x\bar{y} = 0$$

and

$$\bar{x}y = \overline{x + y}$$

A seriatim examination of the possibilities shows that $(B;0,1)$ becomes a $\underline{T}'$-algebra with $1 = \bar{0}$. Now the ideals of A are just the hereditary sub-semigroups of $(A;+,\le,0)$ and so are in correspondence with the ℓ-ideals of G, [4; Theorem 1.2]. Moreover, it is not hard to show that A is an ideal of B, and the ideal-lattice of B is obtained by adjoining a new largest element to the ideal lattice of A. Thus, B is a subdirectly irreducible $\underline{T}'$-algebra, which is never simple, when and only when G is a subdirectly irreducible abelian ℓ-group. When G is simple, so that each positive element is a strong unit and so, by Hölder's Theorem (see Fuchs [10; Theorem 1, p. 45]), is isomorphic to a subgroup of the additive group of the real numbers with the natural ordering, B is subdirectly irreducible with only three ideals.

Given an o-group (totally ordered ℓ-group) H, one can produce a subdirectly irreducible group G simply by taking G as the lexicographic product ([10; p. 25]) HS of H with any simple abelian ℓ-group S. Now the abelian o-groups form a proper class: take lexicographic products (Hahn products) of the additive group of reals over index sets which correspond to the ordinals. Combining these observations, we obtain:

THEOREM 7.3. The variety $\underline{T}'$ is not residually small.

It is possible to give an abstract characterization of the $\underline{T}'$-algebra B which was constructed above; we will give the details on another occasion.

Before closing this section, the author would like to thank Gary Davis and Professor Laszlo Fuchs for helpful discussions on ℓ-groups.

8.   FINITE COMMUTATIVE ALGEBRAS SATISFYING (L)

A well known (see [10; Example 4, p. 16]) and elegant example of an abelian ℓ-group is supplied by the multiplicative group G of positive rationals with the positive cone G^{+} taken as the set N of positive integers.   Then, $a \leq b$ in G means that $\frac{b}{a}$ is an integer, i.e. b is divisible by a.   For $m,n \in N$, let (m,n) denote their greatest common divisor, and let $[m,n]$ denote their least common multiple.   Then in G, the positive part of $\frac{m}{n}$ is:

$$(\frac{m}{n})^{+} = \frac{m}{(m,n)} = \frac{[m,n]}{n}$$

Hence, using the construction of 3.6, the set N of positive integers becomes a commutative BCK-algebra, m which the BCK-product is

$$m*n = \frac{m}{(m,n)} = \frac{[m,n]}{n}$$

the distinguished element is 1, and N has the divisibility-ordering.   Of course, the infimum and supremum of $m,n \in N$ are (m,n) and $[m,n]$, respectively. Moreover, N, and so its subalgebras, satisfies (L): $(\frac{m}{(m,n)}, \frac{n}{(m,n)}) = 1$. Here are two examples of subalgebras of $(N;*,1)$.

EXAMPLE 8.1.   Let m be a fixed positive integer and $D_{m} = \{a \in N: 1 \leq a \leq m\}$, where $\leq$ is the usual order and not the divisibility-order, on N.   For $a,b \in D_{m}$, certainly $\frac{a}{(a,b)} \leq m$.   Hence, $(D_{m};*,1)$ is a subalgebra of $(N;*,1)$ and so a commutative BCK-algebra satisfying (L).   It also follows from the remarks made in Section 5 that D_{m}, ordered by divisibility, is a nearlattice in which each principal order-ideal is a Kleene algebra.   It is not un-interesting to sketch the Hasse diagrams of D_{6} and D_{15}, say.

EXAMPLE 8.2.   Let m be a fixed positive integer and let N(m) be the principal order-ideal of N generated by m.   Here the order is the divisibility-order and N(m) is nothing but the set of divisors of m.   Then the principal order-ideal N(m) is a subalgebra of $(N;*,1)$ and bounded in its own right.   Hence from the discussion in Section 5, N(m) is a Kleene algebra, where the in-volution is the map $a \to \frac{m}{a}$, $a \in N(m)$; it is well known that N(m) is a Boolean algebra (that is, a bounded implicative BCK-algebra) if and only if m is square-free.

In connection with these two examples, we should note two points. Firstly, A(m) is always a subalgebra of D(m).   Secondly, when $m = p^{n}$ is a prime-power, $A(m)$ is isomorphic, as a BCK-algebra, to the chain A_n^1 of Section 7. More generally, when m has prime factorization $m = p_1^{n_1} \cdots p_k^{n_k}$, and $x \in N(m)$, so that $x = p_1^{n_1(x)} \cdots p_k^{n_k(x)}$ where $n_i(x) \leq n_i$, the map $x \to (a_{n_1(x)}, \ldots, a_{n_k(x)})$ is a BCK-isomorphism of $A(m)$ onto $A_{n_1}^1 \times \ldots \times A_{n_k}^1$. Due to Theorem 7.2 and Lemma 5.6, any finite bounded commutative BCK-algebra looks like such a direct product. Combining these observations, we arrive at our final result:

THEOREM 8.3. A BCK-algebra is isomorphic to a subalgebra of the algebra $D(m)$ of 8.1 for some suitable m if and only if it is finite, commutative and satisfies (L).

REFERENCES

1. W. H. Cornish, A multiplier approach to implicative BCK-algebras, Math. Sem. Notes Kobe Univ., 8 (1980), 157–169.

2. W. H. Cornish, Rings and implicative BCK-algebras, Math. Sem. Notes Kobe Univ., 8 (1980), 147–155.

3. W. H. Cornish, 3-permutability and quasicommutative BCK-algebras, Math. Japon., 25 (1980), 477–496.

4. W. H. Cornish, Lattice-ordered groups and BCK-algebras, Math. Japon., 25 (1980), 471–476.

5. W. H. Cornish, On positive implicative BCK-algebras, Math. Sem. Notes Kobe Univ., 8 (1980), 455–468.

6. W. H. Cornish, Varieties generated by finite BCK-algebras, Bull. Austral. Math. Soc., 22 (1980), 411–430.

7. W. H. Cornish, A 3-distributive 2-based variety, Math. Sem. Notes Kobe Univ., 8 (1980), to appear.

8. W. H. Cornish and R. C. Hickman, Weakly distributive semilattices, Acta Math. Acad. Sci. Hungar., 32 (1978), 5–16.

9. E. Fried, A note on congruence extension property, Acta Sci. Math. (Szeged), 40 (1978), 261–263.

10. L. Fuchs, Partially ordered algebraic systems, Pergamon Press, Oxford, 1963.

11. K. Hałkowska, Congruences, automorphisms and subalgebras of a certain
    Lukasiewicz algebra, Zeszyty Nauk. Wyż. Szkoly Ped. w Opulu Mat., 20
    (1978), 43–55.

12. L. Henkin, An algebraic characterization of quantifiers, Fund. Math.,
    37 (1950), 63–74.

13. R. C. Hickman, Join algebras, Comm. Algebra, 8 (1980), 1653–1685.

14. K. Iséki, An algebra related with a propositional calculus, Proc. Japan
    Acad. Ser. A Math. Sci., 42 (1966), 26–29.

15. K. Iséki, Topics of BCK-algebras, Proceedings of the First Symposium on
    Semigroups, Shimane Univ., 1977, 44–56.

16. K. Iséki, On finite BCK-algebras, Proceedings of the Third Symposium on
    Semigroups, Inter-University House of Kansai, 1979, 11–12.

17. K. Iséki, On finite BCK-algebras, Math. Japon., 25 (1980), 225–229.

18. K. Iséki and S. Tanaka, BCK-algebras, Rev. Roumaine Math. Pures Appl.
    23 (1978), 61–66.

19. K. Iséki and S. Tanaka, Ideal theory of BCK-algebras, Math. Japon., 21
    (1976), 451–466.

20. K. Iséki and S. Tanaka, An introduction to the theory of BCK-algebras,
    Math. Japon., 23 (1978), 1–26.

21. B. Jónsson, Algebras whose congruence lattices are distributive, Math.
    Scand., 21 (1967), 110–121.

22. B. Jónsson, Congruence varieties, Appendix 3, 348–377, in G. Grätzer:
    Universal Algebra, 2nd Edition, Springer-Verlag, New York, 1979.

23. J. Kagan and R. Quackenbush, Monadic algebras, Reports in Math. Logic,
    7 (1976), 53–61.

24. J. A. Kalman, Equational completeness and families of sets closed under
    subtraction, Nederl. Acad. Wetensch. Proc. Ser. A, 63 (1960), 402–406.

25. P. Köhler and D. Pigozzi, Varieties with equationally definable prin-
    cipal congruences, Algebra Universalis, 11 (1980), 213–219.

26. Y. Komori, The separation theorem of the $\aleph_0$-valued Lukasiewicz propo-
    sitional logic, Rep. Fac. Sci. Shizuoka Univ., 12 (1978), 1–5.

27. Y. Komori, Super-Lukasiewicz implicational logics, Nagoya Math. J., 72
    (1978), 127–133.

28. H. Lakser, The structure of pseudocomplemented distributive lattices I.
    Subdirect decomposition, Trans. Amer. Math. Soc., 156 (1971), 335–342.

29. A. S. A. Noor, Isotopes of nearlattices, Ph.D. Thesis, Flinders Univ.,
    1980.

30. R. Padmanabhan and R. W. Quackenbush, Equational theories of algebras
with distributive congruences, Proc. Amer. Math. Soc., 41 (1975),
373-377.

31. M. Pałasiński, Some remarks on BCK-algebras, Math. Sem. Notes Kobe
Univ., 8 (1980), 137-144.

32. A. Romanowska and T. Traczyk, On commutative BCK-algebras, Math. Japon.,
25 (1980).

33. Y. Setō, Some examples of BCK-algebras, Math. Sem. Notes Kobe Univ.,
5 (1977), 397-400.

34. T. Traczyk, On the variety of bounded commutative BCK-algebras, Math.
Japon., 24 (1979), 283-292.

35. H. Yutani, On a system of axioms of a commutative BCK-algebra, Math.
Sem. Notes Kobe Univ., 5 (1977), 255-256.

36. H. Yutani, Quasi-commutative BCK-algebras and congruence relations,
Math. Sem. Notes Kobe Univ., 5 (1977), 469-480.

THE JACOBSON RADICAL OF THE ENDOMORPHISM RING
OF A VALUED VECTOR SPACE

L. FUCHS

Department of Mathematics
Tulane University
New Orleans
Louisiana, U.S.A.

P. SCHULTZ

Department of Mathematics
University of Western Australia
Nedlands, Western Australia
Australia

The application of valued vector spaces to the study of abelian p-groups began about 1955 with the work of Charles [1] and Honda [17]. In the past decade there has been a resurgence of interest in the topic, and valued vector spaces over the field with p elements have been used to classify certain classes of p-groups by their socles; for example, totally projective groups [6], direct sums of torsion-complete groups [15], $p^{\omega+1}$-projective groups [9] and $p^{\omega+1}$-injective groups [21]. Furthermore, when not just the socles but also the higher slices are considered, valued vector spaces can be used to help classify other classes of p-groups; for example almost totally injective groups [10] and [12], $p^{\omega+n}$-projective groups [4], and totally injective groups of non limit length [11]. Surveys of the major results in this area can be found in [7] and [8].

Recently, valued vector spaces have been studied as algebraic objects in their own right [3], [5] and [16]. In particular, we began the study of the endomorphism rings of valued vector spaces in [13]. It turned out that the main theorems of [13] are closely related in content and proof to two well-known theorems of abelian group theory:

(1)   the Baer-Kaplansky Theorem [2, Section 108], which states that
an abelian p-group is determined by its endomorphism ring

(2)   Liebert's Theorem [18] characterizing in ring-theoretical terms
the endomorphism ring of a separable abelian p-group, and its analogue for
homogeneous separable torsion-free groups due to Metelli and Salce [19].

To explain precisely what the corresponding theorems for valued vector
spaces are requires the development of some notation, and will be deferred
to the next section.

This paper concerns another property of endomorphism rings, namely the
Jacobson radical. Once again the results are reminiscent of known theorems
about the Jacobson radicals of endomorphism rings of certain classes of
abelian p-groups.

It was shown by Pierce [20] in 1963 that if G is a torsion-complete
p-group, and J the Jacobson radical of its endomorphism ring, then J consists of all the endomorphisms of G which raise the height of every non-zero
element of the socle of G. The analogous theorem for valued vector spaces
which we prove below is that if V is an injective space, then the Jacobson
radical of its endomorphism ring consists of the value increasing endomorphisms of V.

Our second main result concerns subfree spaces. It is known that the
socle of a totally projective p-group is a subfree valued vector space over
the field with p elements [6]. Hausen showed that if G is a totally projective p-group, then J consists of those endomorphisms of G whose restriction
to the socle are in the nil-radical of the endomorphism ring of this subfree
space [14]. Analogously, we show that if V is a subfree vector space, the
Jacobson radical of its endomorphism ring is the nil-radical.

It should be noted however that our results are not consequences of the
corresponding theorems for abelian groups, nor do they imply these theorems.

SURVEY OF KNOWN RESULTS

Let V be a vector space over a field Φ equipped with a valuation. That is,
there is a totally ordered class Γ with maximum element ∞, in which every
non-empty subset has a supremum, and a function $v: V \to \Gamma$ such that

(1) $v(a) = \infty$ iff $a = 0$

(2) $v(\alpha a) = v(a)$ for all $0 \neq \alpha \in \Phi$ and all $a \in V$

(3) $v(a + b) \geq \min(v(a), v(b))$ for all $a, b \in V$

The range of v is denoted *supp V*, and if Γ is the class of ordinals, $\sup\{v(a) + 1: 0 \neq a \in V\}$ is called the *length* of V.

If V,W are valued vector spaces over Φ, a morphism $\eta: V \to W$ is a Φ-linear transformation which does not decrease values. The set of all endomorphisms of V forms a Φ-algebra, which we denote E(V). If ϕ is a map from V into W, ϕ is called a *pseudo-isomorphism* if:

(1) ϕ is a vector space isomorphism

(2)   for all $a, b \in V$, $v(a) \leq v(b)$ iff $v(\phi a) \leq v(\phi b)$

That is, ϕ and its inverse may not preserve values, but must preserve inequalities between values. It is clear that if $\phi: V \to W$ is a pseudo-isomorphism, then ϕ induces a ring isomorphism of E(V) onto E(W). This result has a converse which is the theorem referred to above as analogous to the Baer-Kaplansky Theorem of abelian group theory:

THEOREM 1.   ([13]) Let V and W be valued vector spaces and $\psi: E(V) \to E(W)$ a ring isomorphism. Then there exists a pseudo-isomorphism $\phi: V \to W$ which induces ψ.

The second result referred to above is a ring theoretic characterization of the endomorphism ring of a valued vector space:

THEOREM 2.   ([13]) For any Φ-algebra E with 1, let I denote the set of primitive idempotents of E. Then there exists a valued vector space V such that E is isomorphic to E(V) iff the following conditions are satisfied:

(1)   For all $\eta \in E$ and $\rho \in I$, there exists $\sigma \in I$ such that $\sigma\eta\rho = \eta\rho$

(2)   For all $\rho, \sigma \in I$, either $\rho E\sigma \neq 0$ or $\sigma E\rho \neq 0$; whichever is non-zero is a 1-dimensional subspace of E

(3)   For all $\eta \in E$ and $\rho, \sigma \in I$, if $\eta E\rho \neq 0$ and $\rho E\sigma \neq 0$, then $\eta E\rho E\sigma \neq 0$

(4)   E is complete and Hausdorff in the topology whose subbase of neighbourhoods of 0 consists of the left annihilators of the elements of I, and the left ideal EI is dense in E in this topology.

THE JACOBSON RADICAL

It will be necessary from now on to assume that Γ is the class of ordinals augmented by a maximum value ∞. Let V be any valued vector space with length λ and denote by J the Jacobson radical of E(V). We have not succeeded in characterizing J for arbitrary V, but we have established upper and lower bounds, which are attained respectively by the injective spaces and the sub-free spaces.

For any valued vector space V of length λ, and for all $\sigma < \lambda$, let

$$V^{\sigma} = \{a \in V: v(a) \geq \sigma\} \text{ and } V_{\sigma} = \{a \in V: v(a) > \sigma\}$$

Let $P = \{\eta \in E(V): \eta \text{ increases values}\}$, let $H = \{\eta \in E(V): \text{for all } \rho \leq \lambda$ there exists $\sigma < \rho$ with $\eta(V^{\sigma}) \subseteq V^{\rho}\}$, and let N be the nil radical of E(V).

It is evident that P and H are two-sided ideals in E(V).

LEMMA 1. For an arbitrary valued vector space V,

$$H \subseteq N \subseteq J \subseteq P$$

Proof. Let $\eta \in H$; there is a decreasing chain of ordinals

$$\lambda = \sigma_0 > \sigma_1 > .. \text{ such that } \eta(V^{\sigma_{m+1}}) \subseteq V^{\sigma_m} \text{ for all } m < \omega.$$

Such a chain must terminate after finitely many steps with $\sigma_k = 0$ for some positive integer k. Since $V^{\sigma_k} = V$ and $V^{\lambda} = 0$, $\eta^k = 0$ so $\eta \in N$.

The inclusion $N \subseteq J$ is well known for any ring.

Now suppose $\eta \notin P$, so there exists $0 \neq a \in V$ with $v(a) = v(\eta a)$. Let W be the subspace of V generated by a and ηa, so $W = \Phi a \oplus \Phi b$, where $v(a) \leq v(b)$ and W is a direct summand of V. Let π be a projection of V onto W, so $\pi J \pi$ is contained in the Jacobson radical of E(W).

If $b = 0$ or $v(a) = v(b)$, then E(W) is a full matrix ring over Φ, so $\pi J \pi = 0$ and hence $\eta \notin J$.

If $v(a) < v(b) < \infty$, there is some non-zero $\alpha \in \Phi$ such that $b = \eta a - \alpha a$, so $(1 - \alpha^{-1}\eta)(a) = -\alpha^{-1}b$ which has value $v(b) > v(a)$. Hence $1 - \alpha^{-1}\eta$ has no left inverse in E(V), so $\eta \notin J$.

We shall say that V has the *Pierce property* if $J = P$, and the *Hausen property* if $J = H = N$. Our next task is to show that injective spaces have the Pierce property, and for this we must recall from [3] and [8] the following structure of injectives:

Any space V of length λ has a basic subspace $B = \bigoplus_{\sigma < \lambda} B(\sigma)$, where $B(\sigma)$ is homogeneous of value σ. There is a canonical embedding θ of V into the direct product $\hat{V} = \prod_{\sigma < \lambda} B(\sigma)$ which is onto iff V is injective. Denote the natural projection $\hat{V} \to B(\sigma)$ by ψ_σ for all $\sigma < \lambda$.

For all ordinals $\sigma < \rho < \lambda$ there are canonical morphisms

$$\pi_{\sigma\rho} : V/V_\rho \to V/V_\sigma$$

such that $\{V/V_\sigma ;\ \pi_{\sigma\rho}\}$ is an inverse system. Let

$$W = \varprojlim \{V/V_\sigma ;\ \pi_{\sigma\rho}\}$$

The canonical morphism $\mu : V \to W$ is an embedding. Now define a map $\phi : W \to \hat{V}$ as follows:

Each $x \in W$ has an expression of the form $x = (\ldots, a_\sigma + V_\sigma, \ldots, a_\rho + V_\rho, \ldots)$ in which $a_\sigma - a_\rho \in V_\sigma$ for each $\sigma < \rho < \lambda$. Put $\phi(x) = (\ldots, \psi_\sigma a_\sigma, \ldots, \psi_\rho a_\rho, \ldots)$. Then ϕ is a well-defined isometry. It follows that V is injective iff $\mu = \phi^{-1}\theta$ is an isometry.

LEMMA 2. Let V be a valued vector space such that $V = V_1 \oplus V_2$, where every value in supp V_2 exceeds every value in supp V_1, and both V_1 and V_2 have the Pierce property.

Then V has the Pierce property.

Proof.   We can view $E(V)$ as the matrix ring

$$\begin{bmatrix} E(V_1) & 0 \\ \mathrm{Hom}(V_1, V_2) & E(V_2) \end{bmatrix}$$

Denote the identity map on V_i by 1_i, for $i = 1, 2$ and let

$$\eta = \begin{bmatrix} \alpha_1 & 0 \\ \beta & \alpha_2 \end{bmatrix} \in P \quad \text{and} \quad \begin{bmatrix} a_1 \\ a_2 \end{bmatrix} \in V_1 \oplus V_2$$

Since

$$\eta \begin{bmatrix} a_1 \\ a_2 \end{bmatrix} = \begin{bmatrix} \alpha_1 a_1 \\ \beta a_1 + \alpha_2 a_2 \end{bmatrix}$$

η increases values exactly if both α_1 and α_2 do. In this case, by hypothesis on V_i, there exist $\gamma_i \in E(V_i)$ such that $(1_i + \gamma_i)(1_i + \alpha_i) = 1_i$, for $i = 1,2$. A morphism $\delta \colon V_1 \to V_2$ satisfies

$$\begin{bmatrix} 1_1 + \gamma_1 & 0 \\ \delta & 1_2 + \gamma_2 \end{bmatrix} \begin{bmatrix} 1_1 + \alpha_1 & 0 \\ \beta & 1_2 + \alpha_2 \end{bmatrix} = \begin{bmatrix} 1_1 & 0 \\ 0 & 1_2 \end{bmatrix}$$

iff $\delta(1_1 + \alpha_1) + (1_2 + \gamma_2)\beta = 0$. Since $1_1 + \alpha_1$ is invertible in $E(V_1)$, such a δ does exist, so $\eta \in J$ as required.

We can now complete the characterization of the Jacobson radical for injective spaces.

THEOREM 3. Let V be an injective valued vector space. Then V has the Pierce property.

Proof.   By Lemma 1, we must show $P \subseteq J$.

Let V have length λ. If V is homogeneous, then $P = J = 0$, so the theorem is true for $\lambda = 1$ and we may assume supp V has at least two values.

If λ is not a limit ordinal, then $V = V_1 \oplus V_2$ where V_1 has length $< \lambda$ and V_2 is homogeneous of value $\lambda - 1$, so the theorem is true by induction and Lemma 2.

Assume then that λ is a limit ordinal and let $\eta \in P$. Since P is an ideal it suffices to show that η is quasi-regular. For all $\sigma < \lambda$, V/V_σ is injective of length $< \lambda$, so has the Pierce property.

Since V is isometric to $\varprojlim\{V/V_\sigma; \pi_{\sigma\rho}\}$ and each V_σ is fully invariant in V, η induces value increasing endomorphisms η_σ of V/V_σ such that, for all $\sigma < \rho < \lambda$, $\pi_{\sigma\rho}\eta_\rho = \eta_\sigma \pi_{\sigma\rho}$. It follows that there exist endomorphisms ξ_σ of V/V_σ such that $(1_\sigma + \xi_\sigma)(1_\sigma + \eta_\sigma) = 1_\sigma$, where 1_σ is the identity map on V/V_σ. Consequently

$$\pi_{\sigma\rho}(1_\rho + \xi_\rho)(1_\rho + \eta_\rho) = (1_\sigma + \xi_\sigma)(1_\sigma + \eta_\sigma)\pi_{\sigma\rho} = (1_\sigma + \xi_\sigma)\pi_{\sigma\rho}(1_\rho + \eta_\rho)$$

Since $1_\rho + \eta_\rho$ is a unit, $\pi_{\sigma\rho}\xi_\rho = \xi_\sigma \pi_{\sigma\rho}$ for all $\sigma < \rho < \lambda$. By the universal property of the inverse limit, these ξ_σ induce an endomorphism ξ of V such that $(1 + \xi)(1 + \eta) = 1$, as required.

We shall now show that subfree spaces have the Hausen property. We refer to [3], [5] and [8] for the definition and properties of subfree, s-closed and s-dense, and for a proof of the following assertion :

V is subfree iff there is a free space F in which V is embedded as an s-dense subspace such that all the values in supp F/V are of cofinality > ω.

LEMMA 3. If V is a countable dimensional s-closed subspace of a free space F, then V is a summand of F.

Proof. It is evident that V can be embedded as an s-closed subspace of a countable dimensional summand H of F. Since H/V is countable dimensional, it is free and hence projective. Therefore V is a summand of H, so a summand of F.

LEMMA 4. If V is a countable dimensional subspace of a subfree space S, then V can be embedded in a countable dimensional free summand W of S.

Proof. First embed S as an s-dense subspace in a free space F such that all values in supp F/S are of cofinality > ω, and let W be an s-closure of V in S. Since supp W = supp V, W is a countable valued subspace of F, so free [3,p30].

If W is not countable dimensional, it has a homogeneous summand B of uncountable dimension, say $B = \underset{\nu<\Omega}{\oplus} \Phi a_\nu$, where Ω is an uncountable ordinal. Since V is s-dense in W, no Φa_ν is orthogonal to V so there exist x_ν in V such that $v(a_\nu - x_\nu) > v(a_\nu)$. Hence each $x_\nu = a_\nu + b_\nu$ for some b_ν with $v(b_\nu) > v(a_\nu)$. But this implies the x_ν are linearly independent, contradicting the countable dimensionality of V.

Next, we show that W is s-closed in F. Let $x \in F \backslash W$, and consider $v(x + W) = \sup\{v(x + y): y \in W\} = \lambda$ say. Since W is s-closed in S and countable dimensional, $v(x + y) < \lambda$ for all $y \in W$ can hold only if $x \notin S$ and λ has cofinality ω. But by the hypothesis on F, $v(x + S) > \lambda$, so $v(x - s) > \lambda$ for some $s \in S$. Then $v(s + W) = v(x + W)$ so by the s-closure of W in S, some $a \in W$ satisfies $v(s + a) = \lambda$. Hence $v(x + a) = \lambda$, as required.

Finally, by Lemma 3, W is a summand of F, so a summand of S.

Following an argument due to Hausen, we now show that a countable dimensional free space has the Hausen property [14].

LEMMA 5.   Let $V = \bigoplus\limits_{n<\omega} \Phi a_n$, where $v(a_n) = \sigma_n$.   Then V has the Hausean property.

Proof.   Let $\eta \in J$; by Lemma 1, η increases values so η fails to lie in H only if $\sup \sigma_n = \lambda$, a limit ordinal and $\eta(V^{\sigma_m}) \neq 0$ for all $m < \omega$.

Assume then that $\eta \notin H$ and choose $b_0 \in V$ with $\eta b_0 = c_0 \neq 0$. By Lemma 1, $v(b_0) < v(c_0)$. Next choose $0 \neq b_1 \in V$ such that $v(b_1)$ exceeds the values of all the components of b_0 and c_0, and such that $\eta b_1 = c_1 \neq 0$.

Continuing in this way, we obtain an infinite sequence $b_0, c_0, b_1, c_1, \ldots$ in V such that V can be decomposed into a direct sum $V = \bigoplus\limits_{k<\omega} W_k$ with $b_k, c_k \in W_k$ such that the values in supp W_{k+1} exceed all the values in supp W_k for all $k < \omega$.

Consequently, there is an endomorphism ξ of V such that $\xi c_k = b_{k+1}$ for all $k < \omega$. Let $\gamma = \xi\eta \in J$, so $\gamma b_k = b_{k+1}$ for every k, and let δ be a left quasi-inverse for γ, so $\gamma = \delta\gamma - \delta$. Then for all $k < \omega$, $b_{k+1} = \delta b_{k+1} - \delta b_k$. The elements $d_n = b_0 + \ldots + b_n$ satisfy

$$d_n = b_0 + \sum_{i=1}^{n} \delta(b_i - b_{i-1}) = (1 - \delta)b_0 + \delta b_n$$

Since δb_n has zero component in $W_0 \oplus \ldots \oplus W_{n-1}$, the component of $(1 - \delta)b_0$ in this summand is $b_0 + \ldots + b_{n-1}$. This holds for every $n < \omega$, so we have a contradiction, proving the lemma.

We can now finally characterize the Jacobson radical for subfree spaces.

THEOREM 4.   Let V be a subfree valued vector space.   Then V has the Hausen property.

Proof.   By Lemma 1, it remains to show that $J \subseteq H$, so suppose not, and let $\eta \in J\backslash H$.

Let V have length λ, so there exists $\rho \leq \lambda$ such that $\eta V^\sigma \nleq V^\rho$ for all $\sigma < \rho$. Choose b_0 such that $\eta b_0 = c_0 \notin V^\rho$. By Lemma 1, $v(b_0) < v(c_0)$. Next choose $b_1 \notin V^\rho$ with $v(b_1) \geq v(c_0)$ and $\eta b_1 = c_1 \notin V^\rho$, so $v(b_1) < v(c_1)$.

Continuing in this way, we obtain an infinite sequence $b_0, c_0, b_1, c_1, \ldots$ in V having non-decreasing values and with $v(b_i) > v(b_j)$ whenever $j < i$. By Lemma 4, these elements are contained in a countable dimensional free summand S of V; let ε be a projection of V onto S, so $\varepsilon\eta\varepsilon$ is contained in the Jacobson radical of $E(S)$. Since $\varepsilon\eta\varepsilon b_k = c_k$ for all $k < \omega$, the Hausen property also fails for the space S with $\rho = \sup\{v(b_k): k < \omega\}$.

But this contradicts Lemma 5, so there is no such η.

REFERENCES

1. B. Charles, Etude des groupes abéliens primaires de type $\leq \omega$, Ann. Univ. Sarav. Math.-Natur. Fak. 4 (1955), 184-199.

2. L. Fuchs, Infinite abelian groups, Vol.I and II, Academic Press, New York and London, 1970 and 1973.

3. L. Fuchs, Vector spaces with valuations, J. Algebra 35 (1975), 23-38.

4. L. Fuchs, On $p^{\omega+n}$-projective abelian p-groups, Publ. Math. Debrecen (1976), 309-313.

5. L. Fuchs, Subfree valued vector spaces, Abelian Group Theory, Springer-Verlag Lecture Notes in Mathematics 616 (1977), 158-167.

6. L. Fuchs, The socles of totally projective p-groups, Symp. Math. 23 (1978), 47-62.

7. L. Fuchs, Valued vector spaces and abelian p-groups, Groupe d'Etude d'algèbre (Marie-Paule Malliavin, 2 année : 1976/77) Exp.7, Secretariat Math., Paris, 1978.

8. L. Fuchs, Abelian p-groups and mixed groups, Les Presses de l'Université de Montreal, Montréal, 1980.

9. L. Fuchs and J. M. Irwin, On $p^{\omega+1}$-projective p-groups, Proc. London Math. Soc. 30 (1975), 459-470.

10. L. Fuchs and L. Salce, Almost totally injective p-groups, Quaestiones Math. 1 (1976), 225-234.

11. L. Fuchs and L. Salce, Abelian p-groups of not limit length, Comment. Math. Univ. St. Paul. 26 (1977), 25-33.

12. L. Fuchs and L. Salce, On almost totally injective p-groups and cotorsion groups, J. Algebra 58 (1979), 176-188.

13. L. Fuchs and P. Schultz, Endomorphism rings of valued vector spaces, Rend. Sem. Mat. Univ. Padova, (to appear).

14. J. Hausen, Quasi-regular ideals of some endomorphism rings, Illinois J. Math. 21 (1977), 845-851.

15. P. Hill, A classification of direct sums of closed groups, Acta Math. Acad. Sci. Hungar., 17 (1966), 263-266.

16. P. Hill, Criteria for freeness in groups and valuated vector spaces,
     Abelian Group Theory, Springer-Verlag Lecture Notes in Mathematics 616
     (1977), 140-157.

17. K-Y. Honda, Realism in the theory of abelian groups, I, II and III,
     Comment. Math. Univ. St. Paul. 5 (1956), 37-75; 9 (1961), 11-28; 12
     (1964), 75-111.

18. W. Liebert, Endomorphism rings of abelian p-groups, Studies on Abelian
     groups, Dunod, Paris, 1968, 239-258.

19. C. Metelli and L. Salce, The endomorphism ring of an abelian torsion-
     free homogeneous separable group, Arch. Math. (Basel) 26 (1975), 480-
     485.

20. R. S. Pierce, Homomorphisms of primary abelian groups, Topics in Abe-
     lian groups, Scott, Foresman, Chicago, 1963, 215-310.

21. F. Richman, Extensions of p-bounded groups, Arch. Math. (Basel) 21
     (1970), 449-454.

A SURVEY OF LENGTH PROBLEMS IN GRASSMANN SPACES

J. A. MACDOUGALL[*]

*Mathematics Department
University of Prince Edward Island
Charlottetown, Prince Edward Island
Canada*

1. INTRODUCTION

This survey deals with problems arising in the context of Grassmann products of vector spaces. Background material on Grassman products (exterior powers) can be found in one of the standard references on multilinear algebra, for example [8] or [19]. Throughout, U and V will denote finite-dimensional vector spaces; $\Lambda^r U$ will denote the r-th exterior power of U, and its elements are called r-vectors. The underlying field F will be arbitrary and its elements denoted by Greek letters.

A typical r-vector Z can be expressed in many different ways as a sum of products (also called decomposables or pure r-vectors). If $Z \in \Lambda^r U$ can be written as a sum of k products and no expression for Z has fewer than k products, then Z has *length* k. This is denoted by $\ell(X) = k$. Three problems concerning the length of r-vectors have been studied in recent years, the second and third being closely related:

----

* At the time of writing the author was visiting the Department of Mathematics, The University of Newcastle, N.S.W., Australia.

(1)   What possible lengths may elements of $\Lambda^r U$ have?

(2)   Describe the subspaces of $\Lambda^r U$ whose non-zero r-vectors all have the same length.

(3)   Describe the linear maps from $\Lambda^r U$ to $\Lambda^s V$ which map r-vectors of a fixed length to s-vectors of the same length.

Since there is no general method available for determining the length of an r-vector, these problems are very difficult. We describe the progress that has been made toward their solution.

If $e_1,\ldots,e_n$ is a basis for U, the set $\{e_{i_1} \wedge \ldots \wedge e_{i_r}\}$, where $i_1,\ldots,i_r$ ranges over all increasing r-sequences from $\{1,\ldots,n\}$, forms a basis for $\Lambda^r U$. Thus the dimension of $\Lambda^r U$ is $\binom{n}{r}$. We shall denote a basis vector such as $e_1 \wedge e_2 \wedge e_3$ by E_{123}. Given the expression of any r-vector in terms of some basis, there is a method for determining if its length is 1. The following quadratic equations among the coefficients are called the *Plucker relations or quadratic p-relations* [3, 11]. Here ω ranges over the set of increasing length r sequences chosen from $\{1,\ldots,n\}$.

THEOREM. If $Z = \Sigma \lambda_\omega E_\omega \in \Lambda^r U$, then $\ell(Z) = 1$ if and only if

$$\sum_{i\epsilon\alpha\cup\mu} \varepsilon_{i\alpha\mu} \lambda_{\alpha\backslash i} \lambda_{\mu\cup i} = 0$$

for every pair of sequences α,μ of length $r + 1$ and $r - 1$ respectively where $\varepsilon_{i\alpha\mu}$ is the sign of the permutation which puts $(\alpha\backslash i) \cup (\mu\cup i)$ in increasing order.

There are $n - r$ independent equations in this system. Unfortunately, for all but the smallest values of n and r, these equations are impractical.

It is well known that there is a duality between $\Lambda^r U$ and $\Lambda^{n-r} U$. This is important for our purposes since this isomorphism preserves the length of any r-vector. Thus in most instances, it can be assumed that $r \leq \frac{n}{2}$.

A product $X = x_1 \wedge \ldots \wedge x_r$ determines a unique r-dimensional subspace of U, namely $\langle x_1,\ldots,x_r \rangle$, which we denote by $[X]$. More generally, if $X = X_1 + \ldots + X_k$, then there is a subspace M of U of smallest dimension such that $X \in \Lambda^r M$. This subspace is also denoted $[X]$ and its dimension is called the *rank* of X. Two products X and Y in $\Lambda^r U$ are *adjacent* if $[X]$ and $[Y]$ are adjacent subspaces, i.e. $|[X] \cup [Y]| = r - 1$.

2. POSSIBLE LENGTHS OF r-VECTORS

The space $\Lambda^r U$ contains vectors of length 1 when $0 < r \leq n$, but it is not clear what other lengths vectors in $\Lambda^r U$ may have. Geometers have been interested in the problem since before the turn of the century, mainly in the cases where $r = 2$ or $r = 3$ and n is small. The underlying field naturally was the complex numbers.

If there is an element $Z = \sum_{i \leq k} Z_i$ of length k in $\Lambda^r U$, then $\sum_{i \leq k-1} Z_i$ must be of length $k - 1$, so that we must have vectors of all lengths less than k in $\Lambda^r U$. Thus the problem is reduced to finding the maximum length of any vector $\Lambda^r U$. As in Glassco and Busemann [4], this number will be denoted $N(F,n,r)$ indicating that it is a function of the dimension of U, the size of the product, and the underlying field, F. Glassco [7] showed that the length of an r-vector may depend on the field with the example

$$E_{123} + E_{126} + E_{135} + E_{156} + E_{234} + E_{236} + E_{345} + E_{456}$$

which is length 3 over R but only length 2 over C. Westwick has shown [29] that $N(R,7,3) > N(C,7,3)$ so that the function N does depend on F. It is clear that $N(F_1,n,r) \leq N(F,n,r)$ whenever F_1 is an extension of F. When our results are independent of F, we will write $N(n,r)$.

The value of $N(n,2)$ is easily found. It is based on the fact that the length of $Z = \sum_{i \leq k} u_i \wedge v_i$ is k if and only if $\{u_1,\ldots,u_k,v_1,\ldots,v_k\}$ is an independent set. Thus $2k \leq n$ and so

$$N(n,2) = [\tfrac{n}{2}] \tag{1}$$

Proofs of this result may be found in [7] and [14].

By the length preserving duality, we have

$$N(n, n - 2) = [\tfrac{n}{2}]$$

In fact, for the same reason,

$$N(n,r) = N(n, n - r)$$

so that we need only consider the case when $r \leq \tfrac{1}{2}n$.

For $r > 2$, no explicit formula is known for $N(F,n,r)$ for any fields. However, some upper bounds can be found. Since any r-vector can be expressed as a sum of length 1 basis vectors and the dimension of $\Lambda^r U$ is $\binom{n}{r}$, we have $N(n,r) \leq \binom{n}{r}$. However $N(n,r)$ is always much smaller. Using multilinearity we can collapse sums of basis vectors to get a considerably shorter expression for an r-vector expressed in terms of a basis. For example, when r = 3, careful grouping of the basis vectors yields a sum of at most $(n - 1)^2/4$ products. A somewhat similar approach is taken by Weitzenböck [23]; if $Z = \sum_{i \leq k} Z_i$ has the length k, the Z_i are divided into two classes. For a fixed vector $t \in U$, one class contains those Z_i for which $t \in [Z_i]$ and the other contains the remaining Z_i. We can then write

$$Z = t \wedge \left(\sum_{i \leq s} x_i \wedge y_i \right) + \sum_{i \leq k-s} y_i$$

where $s \leq N(n - 1,2)$ and $k - s \leq N(n - 1,3)$. Replacing $N(n - 1,2)$ with its known value $[\frac{n - 1}{2}]$ yields the recurrence relation:

$$N(n,3) \leq [\frac{n - 2}{2}] + N(n - 1,3)$$

This may be solved to show that

$$N(n,3) \leq \frac{n^2 - 2n - 7}{4}$$

Hutchinson [12] claims that

$$N(R,n,3) \leq \frac{(n - 5)(n + 3)}{4}$$

and that $N(C,n,3)$ is smaller, but no proofs have ever appeared. In any case, all of these suggested upper bounds are quadratic functions of n, whereas the known values of $N(C,n,3)$ indicate a function that is linear rather than quadratic [17].

The approach used above works just as easily for r > 3 and we obtain the more general recurrence (Glassco [4]):

$$N(n,r) \leq N(n - 1,r - 1) + N(n - 1,r) \tag{2}$$

This recurrence yields as an upper bound for $N(n,r)$ a polynomial of the order of $\frac{1}{2r!} n^{r-1}$. The few known values of $N(n,r)$ suggest that this upper bound is not very good.

One simple lower bound can be obtained from the theorem which states that if $\ell(\sum_{i \leq k} Z_i) = k$ and $x \notin [Z_1] + \ldots + [Z_k]$ then $\ell(\sum_{i \leq k} Z_i) = k$. This is proved by Glassco [7] for char $F \neq 0$ and by Lim [14] for any F. Applying this theorem when $Z_i \in \Lambda^2 U$, we get

$$N(n,r) \geq \frac{n - r + 2}{2} \tag{3}$$

The upper and lower bounds (2) and (3) are very far apart.

Several attempts have been made to calculate $N(n,3)$ for small values of n (n = 6,7,8). For n = 6, the bounds given above show that $2 \leq N(6,3) \leq 4$. Schouten in 1931 [22] showed that $N(C,6,3) = 3$ and that for some choice of basis of U, any rank 6 trivector has one of two forms:

(1) $\quad E_{123} + E_{456}$

(2) $\quad E_{123} + E_{145} + E_{246}$

Capdevielle [5] proves that $N(C,6,3) = 3$ and $N(R,6,3) = 3$ and also gives the possible forms a trivector may have. Finally, Glassco shows that $N(F,6,3) = 3$ when char F = 0. Recently, P. Revoy [21] derived canonical forms for 6-dimensional trivectors over any field, but did not discuss their lengths.

When n = 7, inequalities (2) and (3) show that $3 \leq N(7,3) \leq 6$. Glassco [7] proves that $N(F,7,3) \leq 5$ for any field of characteristic 0. This proof involves the quadratic p-relations and is not suitable for studying higher values of n. Both Schouten and Gurevich [9] derived the possible forms a trivector could have over the complex numbers. The rank 7 trivectors are of the following forms:

(1) $\quad E_{123} + E_{453} + E_{673} + E_{257} + E_{164}$

(2) $\quad E_{123} + E_{456} + E_{257} + E_{376}$

(3) $\quad E_{123} + E_{453} + E_{256} + E_{763}$

(4) $\quad E_{123} + E_{456} + E_{257}$

(5) $\quad E_{123} + E_{453} + E_{673}$

Only the first form has more than 4 products and the following example [17] shows that it has length less than or equal to 4.

$$E_{123} + E_{453} + E_{673} + E_{257} + E_{164} = (e_1 - e_5) \wedge (e_2 + e_4) \wedge e_3 + E_{673}$$
$$+ e_2 \wedge e_5 \wedge (e_7 - e_3) - e_1 \wedge e_4 \wedge (e_6 + e_3)$$

Westwick [28] has shown that forms (1) and (3) indeed have length 4, so that

$$N(C,7,3) = 4$$

In the case of n = 8, the upper and lower bounds show

$$4 \leq N(8,3) \leq 8$$

Gurevich [10] again has derived the possible forms that a complex rank 8 trivector may have, showing that $N(C,8,3) \leq 7$. The forms are:

$$(1) \quad E_{123} + E_{456} + E_{147} + E_{267} + E_{357} + E_{258} + E_{368}$$
$$(2) \quad E_{123} + E_{456} + E_{147} + E_{267} + E_{258} + E_{368}$$
$$(3) \quad E_{123} + E_{456} + E_{147} + E_{258} + E_{368}$$
$$(4) \quad E_{147} + E_{456} + E_{267} + E_{357} + E_{258} + E_{368}$$
$$(5) \quad E_{147} + E_{267} + E_{357} + E_{258} + E_{368}$$
$$(6) \quad E_{456} + E_{147} + E_{267} + E_{258} + E_{368}$$
$$(7) \quad E_{456} + E_{147} + E_{258} + E_{368}$$
$$(8) \quad E_{147} + E_{258} + E_{368}$$
$$(9) \quad E_{123} + E_{456} + E_{147} + E_{267} + E_{357} + E_{368}$$
$$(10) \quad E_{123} + E_{456} + E_{147} + E_{267} + E_{368}$$
$$(11) \quad E_{123} + E_{456} + E_{147} + E_{368}$$
$$(12) \quad E_{456} + E_{147} + E_{267} + E_{357} + E_{368}$$
$$(13) \quad E_{147} + E_{267} + E_{357} + E_{368}$$

It is not hard to see that all of these forms can be reduced to a sum of at most 5 products [21]. Westwick has shown that form (9) has length 5 [28], so that

$$N(C,8,3) = 5$$

In addition, he has shown that all other classes have length 4 or less, so that $N(C,8,3)$ is "almost" 4.

No exact results are known for any other fields when $n = 8$. When $r = 3$ and $n > 8$ or when $r > 3$, no exact results are known.

3. SUBSPACES OF VECTORS OF THE SAME LENGTH

There would not seem to be any hope of completely describing subspaces whose vectors all have the same length until the problem of determining the length of a vector has been solved. Therefore progress has essentially been restricted to the simplest cases of length 1 or $r = 2$.

We deal first with the subspaces of $\Lambda^r U$ containing only length 1 vectors. These have been completely characterized and appear to be first described by Westwick [24]. Their description is based on the following nice result.

PROPOSITION. If $\ell(X) = \ell(Y) = 1$ in $\Lambda^r U$, then $\ell(X + Y) = 1$ if and only if $[X]$ and $[Y]$ are adjacent.

Thus the problem is reduced to finding all possible collections of pairwise adjacent r-dimensional subspaces of U. But these have long been known by geometers to fall into only 2 classes: a collection of r-dimensional subspaces of an $(r + 1)$-dimensional space and a collection of r-dimensional subspaces having a common $(r - 1)$-dimensional intersection [2]. Thus we have:

THEOREM. A length 1 subspace M of $\Lambda^r U$ is one of 2 possible types:

   (1) $M \subseteq x_1 \wedge \ldots \wedge x_{r-1} \wedge U$

   (2) $M \subseteq \Lambda^r U_1$ $|U_1| = r + 1$

We note that the dimension of a length 1 subspace is at most $n - r + 1$ if it is of the first type, and $r + 1$ if it is of the second type. Detailed proofs of these results are given in [18].

A knowledge of the simple geometric structure of the length 1 subspaces of $\Lambda^r U$ plays an important role in the description of the linear maps on $\Lambda^r U$ which map length 1 vectors to length 1 vectors. It might be hoped then that there is a reasonably simple characterization of length k subspaces which can be used in an analysis of length k preserving linear maps. From what little is known, however, this does not appear to be the case.

The case $r = 2$ is of special interest because of an isomorphism of $\Lambda^2 U$ with H_n, the space of $n \times n$ skew-symmetrical matrices where $n = |U|$. The isomorphism is described by Marcus and Westwick [20] as follows: let $u_1, \ldots, u_n$ be a basis of U and define $\phi: \Lambda^2 U \to H_n$ on the basis by $\phi(u_i \wedge u_j) = E_{ij} - E_{ji}$ where E_{ij} has 1 in position (i,j) and 0 elsewhere. The vector Z has rank k if and only if $\phi(Z)$ is a length 2k matrix.

Only in the case of $\Lambda^2 U$ have subspaces of any other length been completely described and even then only for length 2 and F a quadratically closed field. This was done by M.J.S. Lim [15]. Certainly any length 2 vector generates a 1-dimensional subspace. When $|U| = 4$ it is shown by Lim [14] that these are the only length 2 subspaces. She first proves that any vectors in H can be written in the form $x_1 \wedge u + x_3 \wedge v$ for some fixed x_1, x_3.

Then supposing $Z_1 = x_1 \wedge x_2 + x_3 \wedge x_4$ and $Z_2 = x_1 \wedge u + x_3 \wedge v$ are in H, a non-trivial linear combination of Z_1 and Z_2 is written as

$$\lambda Z_1 + Z_2 = x_1 \wedge (\lambda x_2 + b_2 x_2 + b_3 x_3 + b_4 x_4) + x_3 \wedge (\lambda x_4 + d_2 x_2 + d_4 x_4)$$

where $u = \sum_2^4 b_i x_i$ and $v = \sum_2^4 d_i x_i$. This has length 2 if and only if the 4 vectors are independent, i.e.

$$\det \begin{vmatrix} 1 & 0 & 0 & 0 \\ 0 & \lambda + b_2 & b_3 & b_4 \\ 0 & 0 & 1 & 0 \\ 0 & d_2 & 0 & \lambda + d_4 \end{vmatrix} \neq 0$$

This is a non-zero 2nd degree polynomial $f(\lambda)$, so that $\ell(\lambda Z_1 + Z_2) = 1$ if and only if F is quadratically closed.

A subspace of the type constructed above is called a $(1,1)$-subspace. It is easy to see that if the u_i are independent the vectors

$$f_i = u_1 \wedge u_i + u_2 \wedge u_{i+1} \qquad i = 3,\ldots,n-1$$

form a basis for a length 2 subspace. Thus $(1,1)$-subspaces of all dimensions up to $n - 3$ exist for every $n \geq 4$. Furthermore, it is shown that all length 2 subspaces of dimension 2 or dimensions greater than 3 must be of this type. Unfortunately, exceptions occur for dimension 3 and these cannot be so concisely described. For example, when $|U| = 5$, the following independent vectors generate a length 2 subspace:

$$\begin{aligned} Z_1 &= u_4 \wedge u_1 + u_3 \wedge u_2 \\ Z_2 &= u_5 \wedge u_2 + u_3 \wedge u_1 \\ Z_3 &= (u_4 + u_5) \wedge u_3 + u_2 \wedge u_1 \end{aligned} \qquad (4)$$

Again, the assumption is made that F is quadratically closed.

Over other fields, there are other subspaces. For example, the vectors $u_1 \wedge u_3 + u_2 \wedge u_4$ and $u_1 \wedge u_4 - u_2 \wedge u_3$ span a 2-dimensional length 2 subspace of $\Lambda^2 U$ ($|U| = 4$) over the reals. This can be easily seen by an argument like that above. In fact, Westwick [25] produces a 3-dimensional length 2 subspace with basis

$$X_1 = u_1 \wedge (\alpha_5 u_2 + \alpha_1 u_4) + u_2 \wedge u_3$$
$$X_2 = u_1 \wedge (\alpha_4 u_4 - \alpha_2 u_3) + u_2 \wedge u_4$$
$$X_3 = u_1 \wedge (\alpha_3 u_2 - \alpha_6 u_3) + u_3 \wedge u_4$$

By using the quadratic p-relations, he shows this is 3-dimensional if and only if there are $\alpha_i \in F$, $1 \leq i \leq 6$, such that the only solution to

$$\alpha_1 \lambda_1^2 + \alpha_2 \lambda_2^2 + \alpha_3 \lambda_3^2 + \alpha_4 \lambda_1 \lambda_2 + \alpha_5 \lambda_1 \lambda_3 + \alpha_6 \lambda_2 \lambda_3 = 0$$

in F is $\lambda_1 = \lambda_2 = \lambda_3 = 0$ ($\alpha_1, \alpha_2, \alpha_3$ are non-zero). These fields include, for example, the real numbers.

Even in $\Lambda^2 U$, length k subspaces for k > 2 have not been completely char-acterized. However, some interesting examples exist. One can generalize the construction for (1,1)-subspaces above. For example, if the u_i are all inde-pendent, define Z_i, $i = k + 1, \ldots, n - k + 1$ as follows:

$$Z_i = u_1 \wedge u_i + u_2 \wedge u_{i+1} + \ldots + u_k \wedge u_{i+k-1} \tag{5}$$

Then the Z_i form a basis for a length k subspace of $\Lambda^2 U$ of dimension n − 2k + 1. Subspaces of this kind may be considered generalizations of length 1 subspaces of the form u ∧ U. Unfortunately, as in the case length 2, these are not the only type of length k subspaces. For example, if U is of dimension 9 and W is a length 2 subspace of U with a basis as in (4), then the vectors

$$Z_1' = u_6 \wedge u_7 + Z_1$$
$$Z_2' = u_6 \wedge u_8 + Z_2$$
$$Z_3' = u_6 \wedge u_8 + Z_3$$

form a basis for a length 3 subspace of U.

This sort of construction can be used to produce subspaces of fixed rank in $\Lambda^r U$ for r > 2. We recall that if $X \in \Lambda^p U$, then u ∧ X has the same length as X provided that $u \notin [X]$. Consequently, if $\{Z_1, \ldots, Z_t\}$ is a basis for a length k subspace of $\Lambda^2 U$ as in (5) and $u \notin \langle [Z_1], \ldots, [Z_t] \rangle$, then $\{u \wedge Z_1, \ldots, u \wedge Z_t\}$ is a basis for a length k subspace of $\Lambda^3 U$. Repeating the process r − 2 times produces length k subspaces of $\Lambda^r U$. Naturally this can be done only if there are r − 2 independent vectors of U outside of

$\langle[Z_1],\ldots,[Z_t]\rangle$. Thus the maximum dimension of this type of length k subspace in $\Lambda^r U$ is $n - 2k - r + 3$. There are other types of length k subspaces however. Letting $\{Z_1, Z_2, Z_3\}$ be as in (4), for example, multiplying by independent vectors outside $\langle[Z_1],[Z_2],[Z_3]\rangle$ produces exceptional three-dimensional length 2 subspaces of $\Lambda^r U$ for all $r \leq n - 3$.

The study of length 2 subspaces of $\Lambda^2 U$ involved a tedious case-by-case analysis. This and the many exceptional examples indicate that it will be difficult to determine all length k subspaces in $\Lambda^2 U$. When $r > 2$, at the very least there will be exceptional examples of length 2 subspaces of dimension $r + 1$. Even if one could determine the length of an r-vector, the problem seems very difficult.

Using the isomorphism of $\Lambda^2 U$ with H_n mentioned above, we are able to describe the length 2k subspaces of H_n. For example, any rank 4 subspace H of H_n is isomorphic under ϕ^{-1} to a length 2 subspace S of $\Lambda^2 U$. If dimension $H \neq 3$, these are all of type (1,1) so a basis for U can be chosen enabling us to write S as

$$S = \langle u_1 \wedge u_3 + u_2 \wedge u_4, u_1 \wedge u_4 + u_2 \wedge u_5 \rangle = \langle Z_1, Z_2 \rangle$$

Under ϕ the element $\alpha Z_1 + \beta Z_2$ of S is mapped to the rank 4 matrix

$$
\left[
\begin{array}{cc|ccc|c}
 & & \alpha & \beta & 0 & \\
\multicolumn{2}{c|}{0} & 0 & \alpha & \beta & 0 \\
\hline
-\alpha & 0 & & & & \\
-\beta & -\alpha & \multicolumn{3}{c|}{0} & 0 \\
0 & \beta & & & & \\
\hline
\multicolumn{2}{c|}{0} & \multicolumn{3}{c|}{0} & 0
\end{array}
\right]
\tag{6}
$$

Thus there is a non-singular matrix P such that for every matrix A in H, $P^{-1}AP$ has the form (6) for some α and β. Similarly, we can find the rank 2k subspaces of H_n corresponding to the length k subspaces of $\Lambda^2 U$ generated by elements of the type (5). Letting $i = k + 1$ to get the image of a typical basis vector, we find

$$\phi(Z_{k+1}) = \begin{bmatrix} 0 & I_k & 0 \\ -I_k & & \\ 0 & & 0 \end{bmatrix}$$

Then an arbitrary element $Z = \sum_{k+1}^{n-k+1} \alpha_i Z_i$ is mapped under ϕ to the rank $2k$ matrix

$$\phi(Z) = \begin{bmatrix} & & \alpha_{k+1}\ \alpha_{k+2} \cdots \alpha_{n-k+1} \\ 0 & & \\ & & \alpha_{k+1}\ \alpha_{k+2} \cdots \alpha_{n-k+1} \\ \hline -\alpha_{k+1} & & \\ \quad -\alpha_{k+1} & & \\ -\alpha_{n-k+1} & & 0 \\ \quad -\alpha_{n-k+1} & & \end{bmatrix}$$

and the set of matrices of this form comprise a rank $2k$ subspace of H_n of dimension $n - 2k + 1$.

In addition, if dimension $H = 3$, we find the rank 4 subspaces of H_n corresponding to the exceptional length 2 subspaces of $\Lambda^2 U$.

4. LINEAR MAPS PRESERVING A FIXED LENGTH

In this section we discuss the problem of characterizing linear maps from $\Lambda^r U$ to $\Lambda^s V$ which for some fixed k send length k vectors to length k vectors. For simplicity, we call these *length k preservers*. It was noted in the introduction that the isomorphism between $\Lambda^r U$ and $\Lambda^{n-r} U$ preserves all lengths. There is a second example of a map which preserves all lengths. For any linear map $f: U \to V$, we define $\Lambda^r f: \Lambda^r U \to \Lambda^r V$ by $\Lambda^r f(x_1 \wedge \ldots \wedge x_r) = f(x_1) \wedge \ldots \wedge f(x_r)$. This is the map said to be induced by f and is injective if and only if f is. For any length k r-vector Z, we clearly have $\ell\left(\Lambda^r f(Z)\right) \leq \ell(Z)$. Then if f is injective, $\ell\left(\Lambda^r f^{-1}(\Lambda^r f(Z))\right) \leq \ell(Z)$ also, so that $\Lambda^r f$ in fact preserves all lengths. These two kinds of linear maps can be thought of respectively as correlations and collineations on the vector space U. Let us call these basic maps.

The study of length k preservers was begun by Westwick for the case k = 1. There was hope of solution here because one can identify the length 1 vectors by means of the quadratic p-relations. In a series of papers he has completely solved this case [24,25,26]. The proofs depend heavily on the characterization of length 1 subspaces described earlier and involve long case-by-case arguments.

In [24], Westwick showed that if T: $\Lambda^r U \to \Lambda^r U$ is a length 1 preserver over an algebraically closed field, then T is injective. Further, if $n \neq 2r$, T must be an induced map as described above. If n = 2r, T might be a composition of an induced map and a correlation. If F is not algebraically closed, then T need not be injective. In [25], there is an example of a singular linear map on $\Lambda^r U$ which preserves length 1. Whether this is essentially the only example is not known. However, if T is a singular length 1 preserver, we have the surprising result that $T(\Lambda^r U)$ is a length 1 subspace; i.e. the image of every vector has length 1 or 0.

Finally, more general situations are studied in [26]. Maps from $\Lambda^r U$ to $\Lambda^s V$ are considered where $r \neq s$ and $|U| \neq |V|$. Here we find both interior and exterior multiplication as length preserving maps under certain conditions. It is surprising to discover, however, that these both can be expressed as compositions of basic maps. Westwick proves

THEOREM. If T: $\Lambda^r U \to \Lambda^s V$ is a length 1 preserver, then either

    (1) $T(\Lambda^r U)$ is a length 1 subspace of $\Lambda^s V$

or

    (2) T is injective and a composition of basic maps.

In fact, he is able to characterize T more precisely as M ∘ P ∘ B where B is a composition of injective basic maps, P is induced by the inclusion of a subspace and M is an exterior multiplication.

If the condition of preserving length 1 is relaxed to allow length 1 vectors to be sent to 0 as well, then we call T a *pure map*. Using the same kind of analysis of the behaviour of length 1 subspaces, Westwick has described the pure maps as well. Of course, T no longer need be injective, but it is surprising that the result is essentially the same.

THEOREM. If $T: \Lambda^r U \to \Lambda^s V$ is a pure map, then either

(1) $T(\Lambda^r U)$ is a length 1 subspace of $\Lambda^s V$

or

(2) T is a composition of basic maps.

Again, T is described precisely as $M \circ P \circ B \circ Q \circ D$ where M, P and B are as before, Q is induced by a projection and D is an interior multiplication.

Length 2 preservers on $\Lambda^2 U$ were investigated by M.J.S. Lim [16] using the knowledge of length 2 subspaces obtained in [15] described previously. She was able to prove that a linear map T is a length 2 preserver if and only if it is a length 1 preserver and F is algebraically closed. Except for n = 4, T then must be an induced map. However, as the difficulty of determining the length k subspaces seems to increase rapidly with k, this method of finding length k preservers is not useful in general. M.H. Lim in [13] has made use of the isomorphism between $\Lambda^2 U$ and H_n, the space of skew-symmetric matrices over F, to prove for any k, if an injective T is a length k preserver then it is a length 1 preserver. Applying the results of Westwick, we see that a singular length k preserver cannot be a length 1 preserver. Lim uses an interesting matrix argument similar to that used for analogous theorems concerning the tensor product (cf. Beasley [1] and Djokovic [6]). This involves studying the characteristic polynomial. His results are valid for any field of at least n elements so long as char $F \neq 2$.

In addition, Lim proves that when $|U| = 2k$, a length k preserver must be injective and hence by the theorem is a length 1 preserver. When $|U| = 2k + 1, k > 2$, a length k preserver T is shown to be a length 1 preserver as long as T does not send any length 1 vectors to 0. These results are true over an algebraically closed field.

The problem of characterizing the length k preservers has been approached by two methods. The approach by M.J.S. Lim is based on having a description of the length k subspaces. This is available only for k = 2 and r = 2 and was obtained on the basis of properties peculiar to $\Lambda^2 U$. It does not appear that this approach will be fruitful for r > 2. The approach taken by M.H. Lim of using the isomorphism with H_n also is limited to r = 2. New methods will be needed to successfully attack this problem.

All these problems discussed here are solved only in cases where special properties hold that do not exist in the general case. It seems likely that until a major breakthrough occurs in the basic problem of determining the length of an r-vector, no substantial progress will be made on the last two closely related problems. It would appear more hopeful that better bounds on the maximum length of an r-vector can be achieved even though an exact solution for all fields may be difficult.

ACKNOWLEDGEMENTS

This paper was prepared while the author was a visitor at the University of Newcastle, Australia. I would like to thank the Mathematics Department there for its hospitality during my visit.

I would also like to thank Dr. L. J. Cummings, the supervisor of my M. Phil. thesis, on which this paper is based.

REFERENCES

1. L. B. Beasley, Linear transformations on matrices: the invariance of rank k matrices, Linear Algebra Appl., 3 (1970), 407–427.

2. E. Bertini, Einfuhrüng in die Projektive Geometrie Mehrdimensionaler Raume, Siedel and Sohn, Vienna, 1924.

3. N. Bourbaki, Algebra I, Chaps 1–3, Addison-Wesley, Reading, 1974.

4. H. Busemann and D. E. Glassco, Irreducible Sums of Simple Multivectors, Pacific J. Math., 49 (1973), 13–32.

5. B. Capdevielle, Classification des Formes Trilinéaires Alterneés en Dimension 6, Enseign. Math. (2), 18 (1972), 225–243.

6. D. Z. Djokovic, Linear Transformations of Tensor Products Preserving a Fixed Rank, Pacific J. Math., 30 (1969), 411–414.

7. D. E. Glassco, Irreducible Sums of Simple Multivectors, 1971, Ph. D. thesis, University of Southern California.

8. W. H. Greub, Multilinear Algebra, Springer-Verlag, New York, 1967.

9. G. B. Gurevich, Sur Les Trivecteurs Dans l'Espace à Sept Dimensions, Dokl. Akad. Nauk SSSR, 8–9 (1934), 567–569.

10. G. B. Gurevich, Classification des Trivecteurs Ayant le Rang Huit, Dokl. Akad. Nauk SSSR, 5–6 (1935), 355–356.

11. W. V. D. Hodge and D. Pedoe, Methods of Algebraic Geometry, Vol. 1, Cambridge University Press, Cambridge, 1953.

12. L. C. Hutchinson, Notes on the Numbers of Leaves of an Alternating Tensor, abstract in Notices Amer. Math. Soc., (1949), 277.

13. M. H. Lim, Rank k Preservers on Second Grassmann Spaces, Malaysian J. Sci. 3 (1975), 145-149.

14. M. J. S. Lim, Rank k Grassmann Products, Pacific J. Math., 29 (1969), 367-374.

15. M. J. S. Lim, L-2 Subspaces of Grassmann Product Spaces, Pacific J. Math., 33 (1970), 167-182.

16. M. J. S. Lim, Rank Preservers of Skew-Symmetric Matrices, Pacific J. Math., 35 (1970), 169-174.

17. J. A. MacDougall, A Note on the Length of Trivectors, Canad. Math. Bull., 21 (1978), 371-372.

18. J. A. MacDougall, Rank Problems in Grassmann Products, M. Phil. thesis, University of Waterloo, 1979.

19. M. Marcus, Finite Dimensional Multilinear Algebra, Parts I and II, Marcel Dekker, New York, 1973, 1975.

20. M. Marcus and R. D. Westwick, Linear Maps on Skew-Symmetric Matrices: The invariance of the Elementary Symmetric Functions, Pacific J. Math., 10 (1960), 917-924.

21. P. Revoy, Trivecteurs de Rang 6, in Colloque sur les formes quad-ratiques, Bull. Soc. Math. France Mém., 59 (1979), 141-155.

22. J. A. Schouten, Klassifizierung der Alternierenden Grossen dritten Grades in sieben Dimensionen, Rend. Circ. Mat. Palermo, 55 (1931), 137-156.

23. R. Weitzenbock, Zur Theorie der Komplexgrossen, Monatsh. Math., 48 (1939), 129-140.

24. R. D. Westwick, Linear Transformations on Grassmann Spaces, Pacific J. Math., 14 (1964), 1123-1127.

25. R. D. Westwick, Linear Transformations on Grassmann Spaces, Canad. J. Math., 21 (1969), 414-417.

26. R. D. Westwick, Linear Transformations on Grassmann Spaces III, Linear and Multilinear Algebra, 2 (1974), 257-268.

27. R. D. Westwick, Trivectors in a Space of Seven Dimensions, Canad. Math.
     Bull., 20 (1977), 401.

28. R. D. Westwick, Irreducible Lengths of Trivectors of Rank Seven and
     Eight, Pacific J. Math., 80 (1979), 575-579.

29. R. D. Westwick, private communication, 1980.

SEMIDISTRIBUTIVE LATTICE VARIETIES

HENRY ROSE

School of Mathematical Sciences
The Flinders University of South Australia
Bedford Park, South Australia
Australia

INTRODUCTION

The collection of semidistributive lattice varieties (Definition 1.2.1)
forms an ideal in the lattice of subvarieties. We show that there are eight
infinite sequences of varieties in that ideal with the following property:
each term of every sequence has the next term as its unique join irreducible
cover (Theorem 3.6).

Every variety of each sequence mentioned above is generated by a semidistributive nonmodular subdirectly irreducible lattice which has a unique
critical quotient (Definition 1.2.7). The results concerning lattices with
such properties can be found in 2.1 and 2.2.

Section 1.1 contains a list of known facts about the lattice of subvarieties, and section 1.2 provides some information about free lattices
and introduces such notions as bounded homomorphism, splitting and semidistributivity.

It is shown in 2.1 that if a subdirectly irreducible semidistributive
lattice L does not have sublattices isomorphic to L_{11} and L_{12} (Figure 1)
then L has a unique critical quotient c/a and the principal filter generated

by a and the principal ideal generated by c are distributive. Corollary
2.1.3 establishes a strong connection between pentagons of L and the critical
quotient c/a.

Most of the results of 2.3 are concerned with arbitrary algebras of
finitary type. The characterization of some conjugate varieties (Remark
1.2.4) can be found in this section.

There are some open problems which are suggested by results obtained.
The most difficult ones are concerned with the varieties $\{L_{11}^n\}^V$ and $\{L_{12}^n\}^V$
(Section 4).

Finally we would like to indicate that some standard universal algebraic
and lattice theoretical concepts are used here without defining them ex-
plicitly. For those the reader should consult G. Grätzer [7,8] and P.
Crawley and R. D. Dilworth [2].

1. PRELIMINARIES

1.1. The Lattice Λ

For a given set Σ of lattice identities we denote by Mod (Σ) the class of
those lattices which satisfy every identity in Σ. If a class $\underline{V}$ of lattices
has the property that $\underline{V}$ = Mod (Σ) for some set of identities Σ we shall say
that $\underline{V}$ is a *variety*.

When $\Sigma = \phi$, Mod (Σ) is the class (= variety) of all lattices and hence
includes every other variety. On the other hand, if Σ consists of the
single identity x = y, then Mod (Σ) is the class of all one element lattices
which is included in every other variety. Further if $\{\Sigma_i\}_{i \in I}$ is a family of
sets of identities, then

$$\text{Mod} \left(\bigcup_{i \in I} \Sigma_i \right) = \bigcap_{i \in I} \text{Mod} (\Sigma_i)$$

and so the intersection of a family of varieties is again a variety. It
follows that the varieties form a complete lattice which we shall refer to
as lattice Λ.

THEOREM 1.1.1.   (i) (B. H. Neumann [17]) The lattice Λ is isomorphic to
the dual of the lattice of fully invariant congruences of the free lattice
with countably many generators.

(ii) Λ is a dually algebraic lattice. The dually compact elements of it are exactly the finitely based lattice varieties.

Since lattices are congruence distributive algebras (see N. Funayama and T. Nakayama [5]), we have the following result concerning the lattice Λ.

THEOREM 1.1.2. Λ is a distributive lattice.

NOTATION. For given class K of lattices we denote by

   SK - the class of all sublattices of the members of K

   HK - the class of all homomorphic images of the members of K

   $P_u K$ - the class of all ultraproducts of members of K

   $P_s K$ - the class of all subdirect products of the members of K

   K^V - the variety generated by K

Using congruence distributivity one can derive many very important results for the theory of lattice varieties. These are due to B. Jónsson [10].

THEOREM 1.1.3. Let K be a class of lattices. Then

   (1) Every subdirectly irreducible member of K^V belongs to $HSP_u K$
   (2) $K^V = P_s HSP_u K$

THEOREM 1.1.4. Let K be a finite set of finite lattices. Then

   (1) Every subdirectly irreducible member of K^V belongs to HSK
   (2) $K^V = P_s HSK$

COROLLARY 1.1.5. Under the assumption of Theorem 1.1.4., up to isomorphism K^V has only a finite number of subdirectly irreducible members, and they all are finite. Furthermore, K^V has only a finite number of subvarieties.

THEOREM 1.1.6. (i) Let A and B be finite nonisomorphic subdirectly irreducible lattices. If $|A| < |B|$ then there is an identity which holds in A, but not in B.

(ii) Suppose $\underline{V}_1$ is a lattice variety generated by its finite members, and $\underline{V}_2$ is a subvariety of $\underline{V}_1$. Then $\underline{V}_1$ covers $\underline{V}_2$ if and only if $\underline{V}_1$ contains exactly one subdirectly irreducible lattice that is not in $\underline{V}_2$.

THEOREM 1.1.7. Given two lattice varieties $\underline{V}_1$ and $\underline{V}_2$, every subdirectly irreducible member of $\underline{V}_1 + \underline{V}_2$ belongs to either $\underline{V}_1$ or $\underline{V}_2$.

The next result can be found in R. McKenzie [16].

THEOREM 1.1.8. (i) Every strictly join prime variety is generated by a finite subdirectly irreducible lattice.

(ii) If the variety $\underline{V}$ is generated by a finite subdirectly irreducible lattice, then $\underline{V}$ is a strictly join irreducible member of Λ.

(iii) If $\underline{V}$ is a strictly join irreducible variety, then it can be generated by a single subdirectly irreducible lattice.

(iv) If the variety $\underline{V}$ is generated by a single subdirectly irreducible lattice, then it is join irreducible.

How large is the lattice Λ? The following theorem gives the answer to this question.

THEOREM 1.1.9. (K. Baker [1], R. McKenzie [15] and R. Wille [23]) There are continuously many varieties.

We shall now turn our attention to the description of the structure of the lattice Λ. Since any nontrivial lattice has $\underline{2}$ as a sublattice and since $\underline{2}$ is the unique distributive subdirectly irreducible lattice, we have that the variety of all distributive lattices $\mathcal{D}$ is a unique cover of the variety T of all trivial lattices. Thus, $\mathcal{D}$ is the unique atom of Λ and since any nondistributive variety has a pentagon or diamond as its members, $\mathcal{D}$ has only two covers in Λ.

THEOREM 1.1.10. (G. Grätzer [6] and B. Jónsson [11]) The variety generated by a diamond, $\{M_3\}^V$, has only the two join irreducible covers $\{M_4\}^V$ and $\{M_{3,3}\}^V$ (Figure 1).

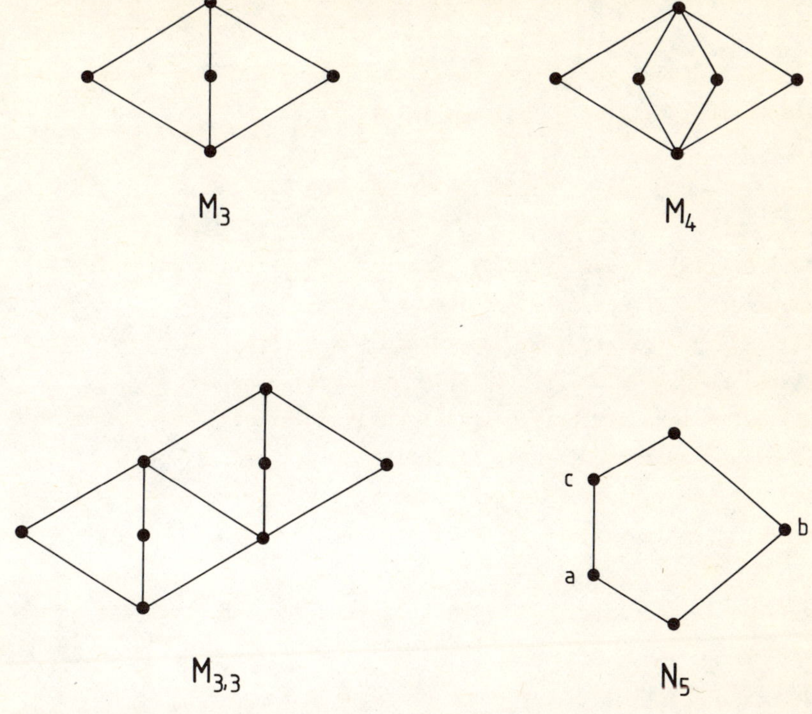

Figure 1

The varieties generated by a pentagon and a diamond both have the same unique join reducible over, namely $\{M_3, N_5\}^V$. R. McKenzie listed fifteen join irreducible covers of a variety generated by pentagon [16]. The next theorem shows that McKenzie's list is complete.

THEOREM 1.1.11. (B. Jónsson and I. Rival [14]) The variety generated by the pentagon, $\{N_5\}^V$, has exactly fifteen join irreducible covers $\{L_1\}^V, \ldots, \{L_{15}\}^V$ (Figure 2(a) and (b)).

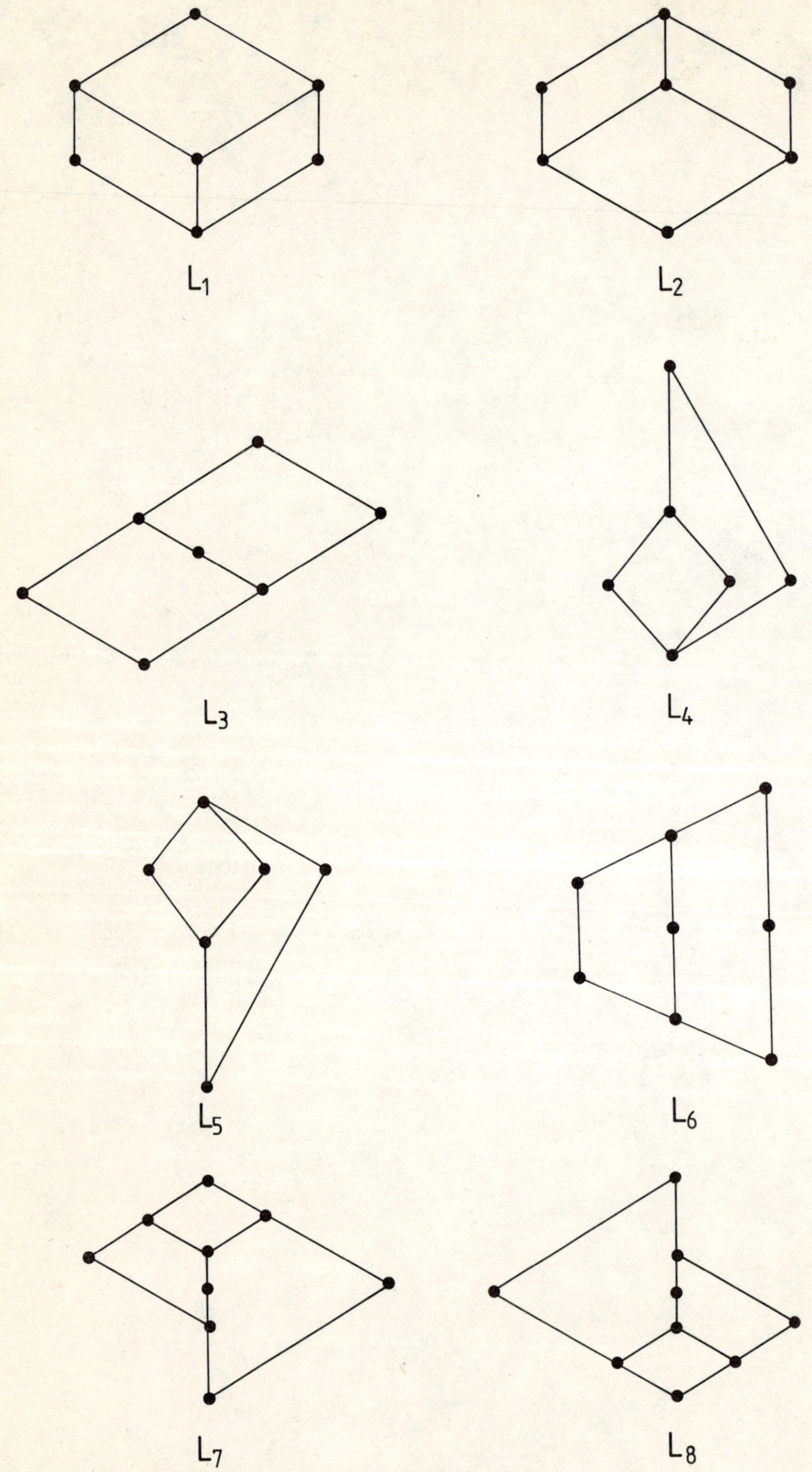

L_1

L_2

L_3

L_4

L_5

L_6

L_7

L_8

Figure 2(a)

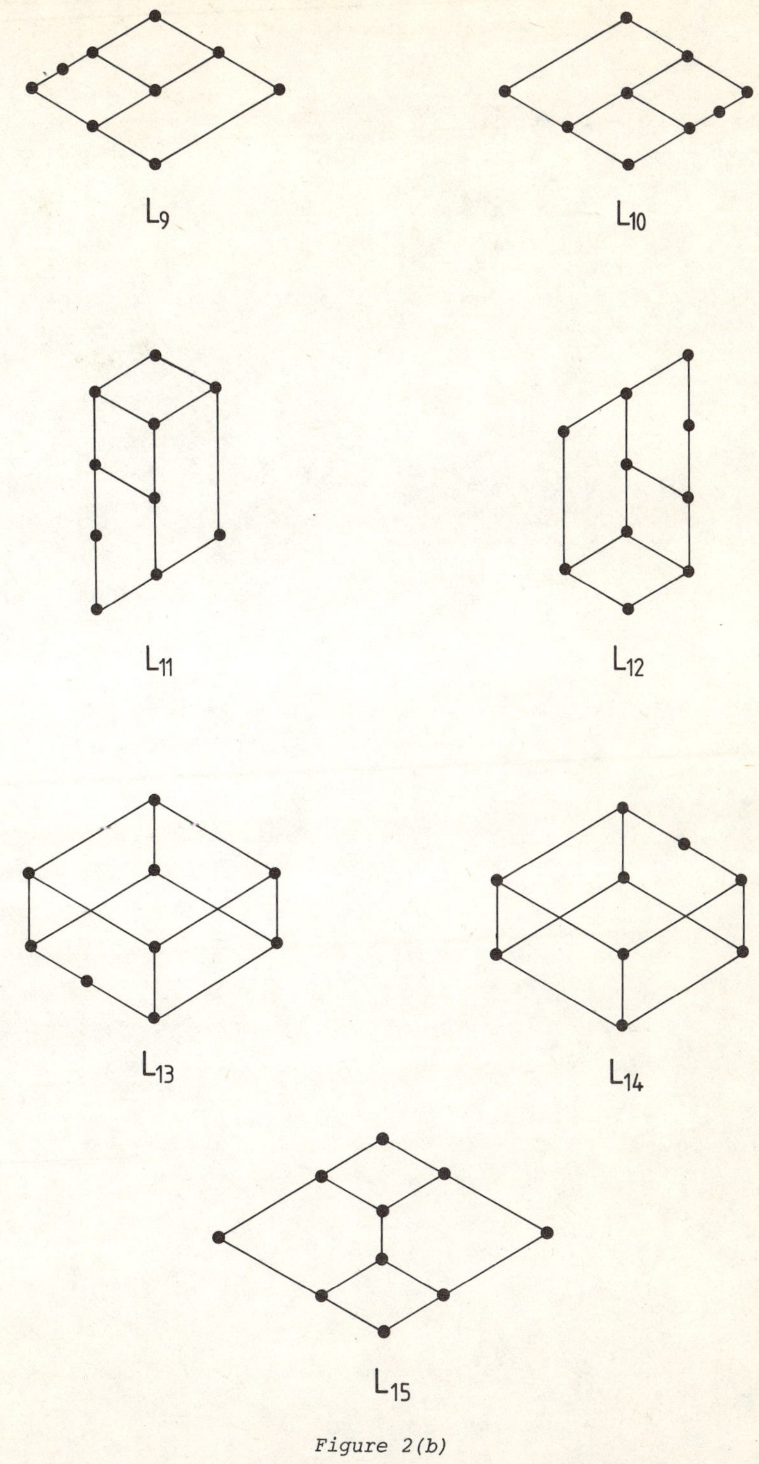

Figure 2(b)

1.2.   Free Lattices, Bounded Homomorphisms, Semidistributivity and the
        Concept of Splitting

DEFINITION 1.2.1.   (i) We shall say that the lattice L satisfies the
Whitman condition if for any a,b,c,d ∈ L we have

   (*W*) ab ≤ c + d if and only if a ≤ c + d or b ≤ c + d
          or ab ≤ c or ab ≤ d

   (ii) We shall say that lattice L is *semidistributive* if for u,a,b,c ∈ L
we have

     (*SD*) u = a + b = a + c implies u = a + bc
   (*SD'*) u = ab = ac implies u = a(b + c)

   A lattice variety $\underline{V}$ is said to be semidistributive if each of its members is a semidistributive lattice.

THEOREM 1.2.2.   (P. M. Whitman [22] and B. Jónsson [9]) Every free lattice
satisfies the Whitman condition and is semidistributive.

DEFINITION 1.2.3.   (R. McKenzie [16])   (i) A pair of lattice varieties is
said to be *splitting* if for every variety $\underline{V}$, either $\underline{V} \subseteq \underline{V}_1$ or $\underline{V}_2 \subseteq \underline{V}$, but
not both.

   (ii)   A lattice epimorphism f: A → B is said to be *bounded* if $f^{-1}(b)$ is
a closed interval for every b ∈ B.   That is, there are order monomorphisms
α,β: B → A (where α is join preserving and β is meet preserving) such that
for all b ∈ B we have $f^{-1}(b) = \beta(b)/\alpha(b)$.

REMARK 1.2.4.   Let $\langle V_1, V_2 \rangle$ be a splitting pair of varieties. It follows
from the definition that $\underline{V}_2$ is a join prime variety, and therefore by Theorem 1.1.8(i), is generated by a finite subdirectly irreducible lattice. We
shall refer to such a lattice as a *splitting lattice* and the variety $\underline{V}_1$ as
a *conjugate* variety. It follows from Theorem 1.2.8. below that every conjugate variety is determined by a single identity. We refer to this identity
as a conjugate.

THEOREM 1.2.5.   (R. McKenzie [16]) A finite subdirectly irreducible lattice
is splitting if and only if it is a bounded homomorphic image of a finitely
generated free lattice.

COROLLARY 1.2.6. (R. McKenzie [16]) (i) Bounded homomorphisms preserve semidistributivity and hence every splitting lattice is semidistributive.

(ii) Let $S \neq \underline{2}$ be a splitting lattice. Then $\{S\}^V \supseteq \{N_5\}^V$ and hence S has a pentagon as a sublattice.

DEFINITION 1.2.7. We say that the quotient c/a of a subdirectly irreducible lattice is *critical* if Con (a,c), the smallest congruence relation which identifies a and c, is the unique atom of the congruence lattice of L.

THEOREM 1.2.8. (R. McKenzie [16]) Let c/a be a prime critical quotient of a splitting lattice L. Suppose further that F(n) is a finitely generated free lattice and $f: F(n) \to L$ is an epimorphism bounded below and above by $\alpha, \beta: L \to F(n)$ respectively. Then for any variety $\underline{V}$ we have that $\underline{V} \models \alpha(c) \leq \beta(a)$ if and only if $L \notin \underline{V}$.

The final theorem of this section gives a characterization of semidistributive varieties.

THEOREM 1.2.9. (B. Jónsson and I. Rival [14]) For a given lattice variety $\underline{V}$ the following statements are equivalent:

(1) $\underline{V}$ is semidistributive
(2) $M_3, L_1, \ldots, L_5 \notin \underline{V}$ (see Figures 1 and 2)
(3) For $n = 0, 1, \ldots,$ put $y_0 = y$, $z_0 = z$, $y_{n+1} = y + xz_n$
    and $z_{n+1} = z + x_n$

Then for some natural number m the identity $x(y + z) = xy_m = xz_m$ and its dual hold in $\underline{V}$.

2. SEMIDISTRIBUTIVE SUBDIRECTLY IRREDUCIBLE LATTICES

2.1. Semidistributive Subdirectly Irreducible Lattices with the Unique
    Critical Quotient

The results of this section are taken from H. Rose [19].

We write $N(c/a,b)$ to indicate that the elements a,b,c of a lattice L generate a pentagon N_5 (Figure 1) with c/a as a critical quotient. (We shall refer to this quotient as an N-quotient.) Also we use this symbol to denote the sublattice of L.

If L and L' are lattices, we shall say that L excludes L' if L has no sublattice isomorphic to L'.

THEOREM 2.1.1. Suppose L is a subdirectly irreducible semidistributive lattice that excludes L_{11} and L_{12} (Figure 2(b)). Then L has a critical quotient c/a such that Con (a,c) identifies no two distinct elements of L except a and c. Hence c/a is a prime quotient, a is meet irreducible and c is join irreducible and therefore c/a is the only critical quotient of L.

PROBLEM 2.1.2. Is the converse of the Theorem 2.1.1. true?

COROLLARY 2.1.3. Let L and c/a be as in Theorem 2.1.1. Then for b ϵ L the following conditions are equivalent:

(1) N(u/v,b) for some quotient u/v of L
(2) N(c/a,b)
(3) b is noncomparable with a and c.

COROLLARY 2.1.4. Let L and c/a be as in Theorem 2.1.1. Then the sublattices [a) and (c] are distributive.

COROLLARY 2.1.5. Let L and c/a be as in Theorem 2.1.1., and suppose that u/v is a nontrivial quotient of [c). Then there exists b ϵ L such that N(c/a,b), b $\leq$ u and b $\nleq$ v.

CONJECTURE 2.1.6. Let L be as in Theorem 2.1.1., and suppose that L is generated by a finite set. Then

(1) L is a finite lattice
(2) L is a bounded homomorphic image of a finitely generated free lattice and hence is a splitting lattice (see Theorem 1.2.5).

It was shown in R. McKenzie [16] that every splitting lattice is finite. Thus 2.1.6(2) would imply 2.1.6(1).

Let L and c/a be as in Theorem 2.1.1. Put B = {b ϵ L: N(c/a,b)}. To prove 2.1.6(1) it is enough to show that the set B is finite. Indeed, suppose that $x_0 < x_1 < \ldots$ is an infinite chain in [a). We may assume that $c \underset{\neq}{<} x_0$. By Corollary 2.1.5. this implies that there exists an infinite sequence $b_0, b_1, \ldots$, where for $i = 0, 1, \ldots$, $N(c/a,b)$. Thus if B is a finite set, by Corollary 2.1.5. and its dual the lattices [a) and (c] must be of finite length. Since by Corollary 2.1.4. these two lattices are distributive they must be finite. It follows from Theorem 2.1.1. and Corollary 2.1.2. that $L = [a) \cup (c] \cup B$.

2.2. Splitting Lattices Obtained from Finite Distributive Lattices

The results of this section are taken from H. Rose [18,20].

In [3] and [4], Alan Day introduced a construction that has proved to be extremely useful, the splitting of an interval in a lattice L. Here we need only the special case in which the interval consists of a single element d. In this case the new lattice is the set $L[d] = (L - \{a\}) \cup \{a,c\}$ with $a = (d,0)$ and $c = (d,1)$ and with the partial order defined as follows:

$x \le y$ if and only if either $x,y \in L - \{d\}$ and $x \le y$ in L
     or $x \in \{a,c\}$, $y \in L - \{d\}$ and $d \le y$
     or $x \in L - \{d\}$, $y \in \{a,c\}$ and $x \le d$
     or $(x,y) = (a,a)$, (a,c) or (c,c).

It is easy to check that $L[d]$ is indeed a lattice, and that there is an epimorphism from $L[d]$ to L. In this section we consider the class K of subdirectly irreducible lattices that are obtained from finite distributive lattices by splitting an element. Observe that the pentagon N_5 is in K and also the lattices L_{13}, L_{14} and L_{15} (Figure 2(b)). By Alan Day [4], every lattice in K is splitting.

The next two results are implicitly contained in B. Jónsson and I. Rival [14].

THEOREM 2.2.1. Suppose L is a nondistributive, subdirectly irreducible semidistributive lattice, and c/a is a critical quotient of L. Then the following conditions are equivalent:

    (1) L excludes $L_6, L_7, \ldots, L_{12}$ (Figure 2(a) and (b))
    (2) c/a is the only N-quotient of L
    (3) Con (a,c) identifies no two distinct elements of L except a and c and L/Con (a,c) is distributive
    (4) $L = \mathcal{D}[d]$ for some distributive lattice $\mathcal{D}$ and $d \in \mathcal{D}$

COROLLARY 2.2.2. If the lattice L from the preceding theorem is finitely
generated, then L ϵ K.

The final result of this section completely characterizes the members
of K.

THEOREM 2.2.3. Suppose D is a finite distributive lattice and $0,1 \neq d \epsilon D$.
Then $D[d]$ is subdirectly irreducible if and only if

    (1)   every cover of d is join irreducible

    (2)   every dual cover of d is meet irreducible

    (3)   every prime quotient in D is projective to a prime quotient p/q
with p = d or q = d.

2.3. Projective Subdirectly Irreducible Lattices

DEFINITION 2.3.1. Let $\underline{V}$ be a variety of lattices, and suppose that $P \epsilon \underline{V}$.
We say that P is *projective* in $\underline{V}$ if for any $A,B \epsilon \underline{V}$ and any lattice homomorphisms h: P $\rightarrow$ B and g: A $\rightarrow$ B with g onto there exists a lattice homomorphism f: P $\rightarrow$ A such that h = fg.

We state without proof the following well known result about projective
lattices.

LEMMA 2.3.2. For a variety $\underline{V}$ and lattice $P \epsilon \underline{V}$, the following three conditions are equivalent:

    (1) P is projective in $\underline{V}$.

    (2) For any lattice $M \epsilon \underline{V}$ and any onto homomorphism g: M $\rightarrow$ P there
exists an embedding h: P $\rightarrowtail$ M such that hg is the identity map on P.

    (3) P is a retract of a free lattice.

Our next five results are valid for arbitrary algebras of finitary
type.

LEMMA 2.3.3. If K is a class of algebras and A is a projective subdirectly
irreducible member of K^V, then A isomorphic to a subalgebra of a member of
K.

Proof. Since $A \in K^V$, we have that for some index set I $A \in HS(\Pi A_i)_{i \in I}$, where for any $i \in I, A_i \in K$. Since A is projective, we have $A \in S(A_i)_{i \in I}$ so we may suppose that A is a subalgebra of $(\Pi A_i)_{i \in I}$. Thus A is a subdirect product of the family $\pi_i(A)_{i \in I}$ where each $\pi_i : (\Pi A_i)_{i \in I} \to A_i$ is the natural projection. But since A is subdirectly irreducible, there exists $j \in I$ such that $\pi_j(A) \cong A$ and so $\pi_j : A \to A_j$ is an embedding.

COROLLARY 2.3.4. Let P be a projective subdirectly irreducible algebra. Then for any algebra A, $P \in \{A\}^V$ if and only if P is isomorphic to a subalgebra of A.

COROLLARY 2.3.5. Every projective subdirectly irreducible algebra is splitting.

Proof. Let P be a projective subdirectly irreducible algebra and suppose that $\underline{V}$ is a class of algebras with the following property: $A \in \underline{V}$ if and only if $P \notin \{A\}^V$. We need to show that $\underline{V}$ is a variety. Suppose not, then P belongs to the variety generated by $\underline{V}$. Thus by Lemma 2.3.3. there exists $A \in \underline{V}$ such that P is a subalgebra of A and, therefore, $P \in \{A\}^V$ which leads to a contradiction.

COROLLARY 2.3.6. Let P be a projective subdirectly irreducible algebra and suppose that σ is its conjugate identity. Then $A \in \mathrm{Mod}(\sigma)$ if and only if A does not have a subalgebra isomorphic to P.

Proof. Obviously, if $A \models \sigma$ then A cannot have a subalgebra isomorphic to P. Conversely, suppose that A does not have P as a subalgebra. Then by Corollary 2.3.4. $P \notin \{A\}^V$, and therefore $\{P\}^V \notin \{A\}^V$. Since P is splitting, we conclude that $A \in \mathrm{Mod}(\sigma)$.

COROLLARY 2.3.7. If a variety $\underline{V}$ is generated by its finite members, then the projective subdirectly irreducible algebras of $\underline{V}$ are finite.

Proof. Follows immediately from Lemma 2.3.3.

REMARKS. (i) Since the variety of all lattices is generated by the class of all lattices, it follows from Corollary 2.3.7. that any projective subdirectly irreducible lattice is finite. This can be also obtained by combining results from [16,13].

(ii) The converse of Corollary 2.3.5. is false. Indeed the lattice in Figure 3 fails the Whitman condition and therefore is not embeddable in a free lattice (see Definition 1.2.1. and Theorem 1.2.2.). Thus it is not projective; on the other hand this lattice is splitting.

Next we turn to the problem of characterization of conjugate varieties determined by a self dual identity.

NOTATION. For a given class $\underline{A}$ of lattices we define $\underline{A}_s$ as follows: $L \in \underline{A}_s$ if and only if there exists a lattice $L' \in \underline{A}$ such that $\mathrm{Sub}\,(L) \cong \mathrm{Sub}\,(L')$, that is L and L' have isomorphic lattices of sublattices.

THEOREM 2.3.8. Let P be a self dual projective subdirectly irreducible lattice such that for any lattice P', $\mathrm{Sub}\,(P) \cong \mathrm{Sub}\,(P')$ implies $P \cong P'$. Suppose further that σ is its conjugate identity and $\underline{V} = \mathrm{Mod}\,(\sigma)$. Then $\underline{V} = \underline{V}_s$.

Proof.   Suppose that for some lattice L we have $L \in \underline{V}_s$ and $L \notin \underline{V}$. Then $L \notin \mathrm{Mod}\,(\sigma)$ and since $P \in \{L\}^V$, by Corollary 2.3.4. L includes P. Let $L' \in \underline{V}$ such that $\mathrm{Sub}\,(L') \cong \mathrm{Sub}\,(L)$. Since L includes P the lattice L' must have a sublattice P' such that $\mathrm{Sub}\,(P') \cong \mathrm{Sub}\,(P)$. By assumption $P \cong P'$ and so $P \in \underline{V}$ which leads to a contradiction.

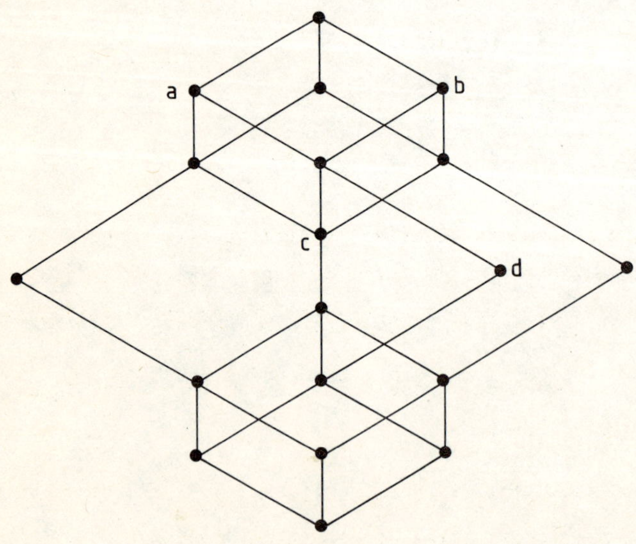

Figure 3

For the definition of cycle see [13].

LEMMA. If a semidistributive lattice L has a cycle, then L has L_{11} or L_{12} as a sublattice.

THEOREM. Let $\underline{V}$ be a semidistributive lattice variety such that $L_{11}, L_{12} \notin \underline{V}$ and suppose that L is a finite lattice in $\underline{V}$. Then

(1) L is a bounded homomorphic image of a finitely generated free lattice.

(2) If L satisfies the Whitman condition, then L is a sublattice of a free lattice (= L is projective).

(3) If L is subdirectly irreducible, then L is a splitting lattice.

CONJECTURE. If a semidistributive lattice L has a cycle, then L has both L_{11} *and* L_{12} as sublattices.

PROBLEM. What conditions must a lattice variety satisfy whose finitely generated subdirectly irreducible members are splitting?

REMARK. It can be shown that the varieties $\{L_{11}^{n}\}_{n<\omega}^{V}$ ($\{L_{12}^{n}\}_{n<\omega}^{V}$) have L_1 (L_2) as members (Figure 2(a)). This implies that if every finitely generated subdirectly irreducible member in the variety $\underline{V}$ is splitting, and $L_i \in \underline{V}$ for i = 11 or i = 12, then there exists $n \geq 0$ such that $L_i^{n} \notin \underline{V}$.

3. THE IDEAL I

The collection of semidistributive lattice varieties which do not have L_{11} and L_{12} (Figure 2) as members forms an ideal I in the lattice Λ. The following two conjectures were suggested by Professor Bjarni Jónsson.

CONJECTURE 3.1. The ideal I has a largest variety which is determined by a single identity.

CONJECTURE 3.2. If $\underline{V} \in I$, then $\underline{V}$ is locally finite.

This conjecture is a stonger version of Conjecture 2.1.6.

The following three problems can be found in [12].

PROBLEM 3.3. Is it true that if the variety $\underline{V}$ has finitely many subvarieties, then $\underline{V}$ is generated by a finite lattice (equivalently, if $\underline{V}$ is generated by a finite lattice, then is the variety that covers $\underline{V}$ also generated by a finite lattice)?

PROBLEM 3.4. Is it true that if the variety $\underline{V}$ is generated by a finite lattice it has only finitely many covers?

PROBLEM 3.5. Is it true that for any natural number m there are only finitely many varieties having height m?

These problems appear to be very difficult, and there is no prospect of
an easy solution. However, the results concerning subdirectly irreducible
members of the varieties belonging to I (see Section 2) gives us hope that
the problems can be solved for those varieties.

There are eight infinite sequences of join-irreducible varieties in I
with the following property: each term of every sequence has the next term
as its unique join irreducible cover.

These are the sequences L_i^n for $i = 6,7,8,9,10,13,14,15$, and $n \geq 0$,
shown in Figure 4(a) and (b); where L_8^n is dual to L_7^n, L_{10}^n is dual to L_9^n, and
L_{14}^n is dual to L_{13}^n. Note that $L_i^0 = L_i$ of Figure 2.

THEOREM 3.6. (H. Rose [21].) For any natural number n, the variety $\{L_i^n\}^V$
for $i = 6,7,8,9,10,13,14,15$, has only one join irreducible cover, namely the
variety $\{L_i^{n+1}\}^V$.

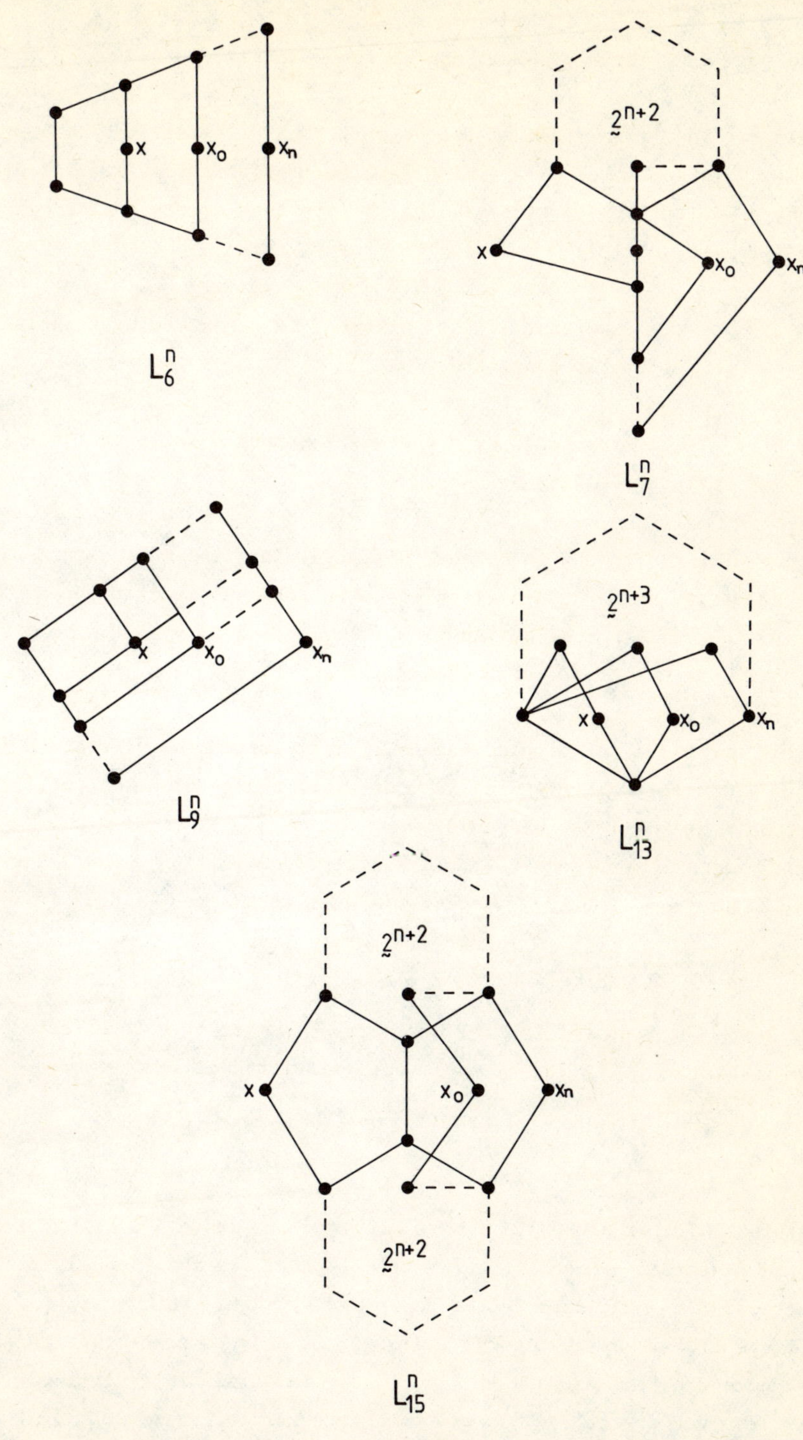

Figure 4(a)

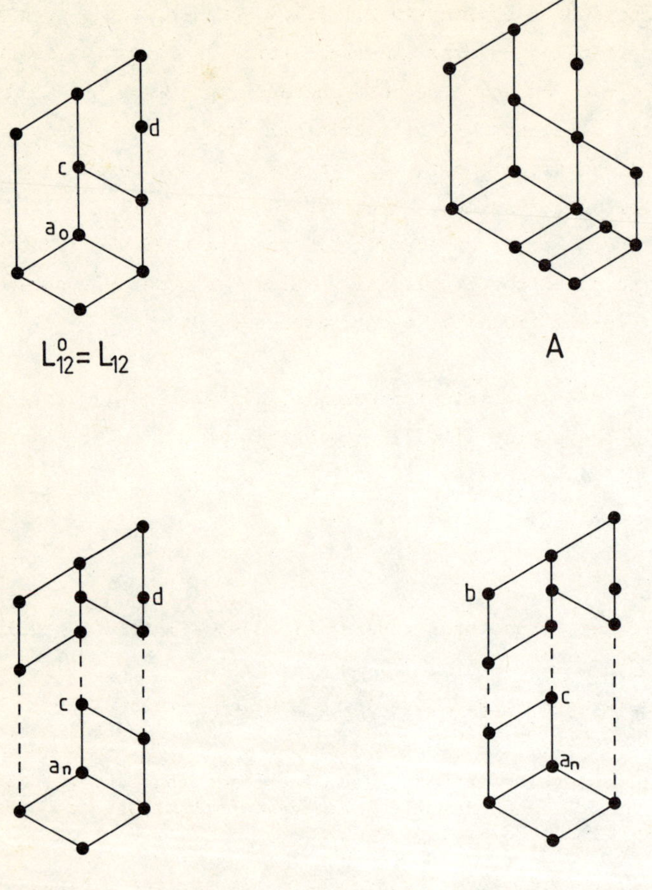

Figure 4(b)

4. THE VARIETIES $\{L_{11}^n\}^V$ AND $\{L_{12}^n\}^V$

The lattices L_{12}^n for n = 0,1,..., are pictured in Figure 4(b). Since for
any n, the lattices L_{11}^n and L_{12}^n are mutually dual, we shall consider only
the lattices L_{12}^n and the varieties generated by them. Note that $L_{12}^0 = L_{12}$
(see Figure 2(b)). Using Theorem 1.1.6. it is easy to check that for any
natural number n, the variety $\{L_{12}^{n+1}\}^V$ is a join irreducible cover of $\{L_{12}^n\}^V$.
However, the variety $\{L_{12}^0\}^V$ has two join irreducible covers: $\{L_{12}^1\}^V$ and $\{A\}^V$.
(For the lattice A see Figure 4(b).)

For any $n = 0,1,\ldots$, the lattice L_{12}^n is semidistributive and has two critical quotients (cf. Theorem 2.1.1.). Further, for n even the quotient $cd/a_n d$ is critical. However, the sublattice $[a_n d)$ is not distributive and the elements $a_n d$, cd and d do not generate a pentagon. Similarly if n is odd, then the quotient $cb/a_n b$ is critical, but the sublattice $[a_n b)$ is not distributive and the elements cb, $a_n b$ and b do not generate a pentagon. Thus Corollaries 2.1.3. and 2.1.4. are not valid for L_{12}^n.

Since in the proof of Theorem 3.6 we have used the results from Section 2 in order to solve the Problem 4.1. below, we need an approach which will be quite different from the one we developed before.

PROBLEM 4.1. How many join irreducible covers does the variety $\{L_{12}^0\}^v$ have? For $n \geq 1$ is the variety $\{L_{12}^{n+1}\}^v$ the unique join irreducible cover of $\{L_{12}^n\}^v$?

REFERENCES

1. K. A. Baker, Equational classes of modular lattices, Pacific J. Math. 28 (1969), 9-15.

2. P. Crawley and R. P. Dilworth, Algebraic Lattice Theory, Prentice-Hall, Englewood Cliffs, N.J., 1973.

3. A. Day, Splitting lattices and congruence modularity, Colloquia Mathematica Societatis János Bolyai 17, Contributions to Universal Algebra, Szeged, (1975), 57-71.

4. A. Day, Splitting lattices generate all lattices, Algebra Universalis 7 (1977), 163-169.

5. N. Funayama and T. Nakayama, On the distributivity of a lattice of lattice-congruences, Proc. Imp. Acad. Tokyo 18 (1942), 553-554.

6. G. Grätzer, Equational classes of lattices, Duke Math. J. 33 (1966), 613-622.

7. G. Grätzer, Universal Algebra, University Series in Higher Mathematics, Van Nostrand, Princeton, N.J., 1968.

8. G. Grätzer, General Lattice Theory, Birkhauser Verlag, Basel and Stuttgart, 1978.

9. B. Jónsson, Sublattices of a free lattice, Canad. J. Math. 13 (1961), 256-264.

10. B. Jónsson, Algebras whose congruence lattices are distributive, Math. Scand. 21 (1967), 110-121.

11. B. Jónsson, Equational classes of lattices, Math. Scand. 22 (1968), 187-196.

12. B. Jónsson, Varieties of lattices: Some open Problems, Colloquia Mathematica Societatis János Bolyai (to appear).

13. B. Jónsson and J. B. Nation, A report on sublattices of a free lattice, Colloquia Mathematica Societatis János Bolyai 17, Contributions to Universal Algebra, Szeged, (1975), 223-257.

14. B. Jónsson and I. Rival, Lattice varieties covering the smallest non-modular variety, Pacific J. Math. 82 (1979), 463-478.

15. R. McKenzie, Equational bases for lattice theories, Math. Scand. 27 (1970), 24-38.

16. R. McKenzie, Equational bases and nonmodular lattice varieties, Trans. Amer. Math. Soc. 174 (1972), 1-43.

17. B. H. Neumann, Universal Algebra, Lecture Notes Courant Inst. of Math. Sci., New York University, 1962.

18. H. Rose, Varieties generated by splitting lattices Q_1, Q_2^n and N_μ, abstract in Notices Amer. Math. Soc. 26 (1979), A-366.

19. H. Rose, Semidistributive subdirectly irreducible lattices with the unique critical quotient, (manuscript).

20. H. Rose, Splitting lattices obtained from finite distributive lattices, (manuscript).

21. H. Rose, The varieties $\{L_i^n\}^V$ for i = 6,7,...,10,13,14,15. (manuscript).

22. P. M. Whitman, Free lattices, Ann. of Math. (2) 42 (1941), 325-330.

23. R. Wille, Primitive subsets of lattices, Algebra Universalis 2 (1972), 95-98.